Silver-Zinc Battery: Best Practices, Facts and Reflections

Silver-Zinc Battery:

Best Practices, Facts and Reflections

Albert Himy

VANTAGE PRESS
New York

FIRST EDITION

Published by Vantage Press, Inc.
516 West 34th Street, New York, New York 10001

Manufactured in the United States of America
ISBN: 0-533-11413-6

Library of Congress Catalog Card No.: 94-90924

0 9 8 7 6 5 4 3 2 1

To my wife, Lucie,
my constant source of energy

Contents

Preface

This book is a sequel to my preliminary book published in 1986, *Silver-Zinc Battery: Phenomena and Design Principles*, which covered fundamental phenomena and basic design principles that should be known first by anybody who intends to deal with the silver-zinc battery system.

The present book covers more detailed manufacturing techniques and sorts out the best practices on the basis of my experience and frequent observations on failed silver-zinc batteries and their performance in the field. Also beside the incontrovertible facts, I added my own reflections and some of the possible explanations to fit the experimental data obtained on actual batteries during their actual usage and life.

I added at various places throughout the book when appropriate and pertinent to the subject some ideas on how to solve or improve the life or at least the reliability of the batteries.

Several features relative to the zinc electrode, the separators, and the electrolyte have a strong commonality with the same components used in other alkaline zinc batteries, the nickel-zinc and the zinc-air (or oxygen) systems. Any improvement made on any zinc system could be adapted to the others. Therefore, the present trend to improve the nickel-zinc and the zinc-air systems for commercial applications will reflect favorably on the silver-zinc system, which is primarily used for military applications and does not have much funding support for improvement. The present concern for silver-zinc batteries is essentially to maintain the current status by monitoring closely the design parameters that worked in the past and rely on the best quality assurance in order to achieve adequate reproducibility and reliability in the field.

I hope that this book will keep the interest of silver-zinc battery designers alive so as to bring their own ideas to this subject.

Silver-Zinc Battery: Best Practices, Facts and Reflections

CHAPTER I
General

It is appropriate first to recall some fundamental features of the silver-zinc system, which were covered in detail in my previous book *Silver-Zinc Battery: Phenomena and Design Principles*.

The silver-zinc system has the highest gravimetric energy density (Wh/kg [Watt-hour/kilogram]) and the highest volumetric energy density (Wh/L [Watt-hour/liter]) of all rechargeable batteries in wide use today. Some rechargeable lithium batteries may have higher values, but they are still in the research stage, small in size, and very often limited in rate capability, besides the fact that safety and environmental concerns are considered an obstacle as to their eventual use.

The silver-zinc battery is currently available from 0.1 Ah (ampere hours) to up to 2500 Ah, mainly for military applications. The battery is not in the research stage (possibly some incremental improvements are in process). It can be designed for low-rate and high-rate capability as required. An application for operation at low rate and requiring very high gravimetric energy density (over 200 Wh/kg) will have short cycle and calendar life. Another application for operation at high rate and requiring long cycle and calendar life had a design with a gravimetric energy density of over 100 Wh/kg. Safety and environmental concerns are not considered problems. In terms of cost, silver is fully and easily recoverable, and in military applications, silver is furnished by the government, so that the cost of the battery includes only inexpensive materials and labor.

The limitations of the silver-zinc battery are its relatively short cycle life (at most 100 cycles) and short calendar life (at best three years). In most applications and in the field, the batteries, not being treated uniformly, last about 50 cycles and, when not too frequently cycled, a maximum of two years. The main causes are separator degradation by the electrolyte and by silver migration, and zinc electrode degradation due to zinc solubility and uneven zinc replating (commonly called shape change).

Description of the Cell

The silver-zinc cell consists of positive electrodes made of silver powder and negative electrodes made of zinc oxide powder. The separator system consists of a combination of porous absorbent material and cellophane layers. The electrolyte is a solution of potassium hydroxide of 40 to 45% concentration. The current collectors or grids in the electrodes are made of silver metal in various forms, and the tabs or leads out of each electrode are made of silver metal.

CHAPTER II
Best Practices

Manufacture

Positive Electrode

The positive electrode (silver) consists of an assembly of a grid or current collector made of silver metal and of very fine silver powder, which, after being pressed to the desired thickness, is sintered at high temperature. This electrode is extremely sturdy and rugged and can be easily handled in production.

Active Material

The active material consists exclusively of pure silver powder (or silver oxide), without any additive.

The silver powder used must have a very fine particle size in the range of 1 to 4 microns. Generally the higher the particle size, the higher the voltage at high rates and the lower the capacity output. This is true also in reverse: the lower the particle size, the lower the voltage and the higher the capacity output.

In general, the 3-micron size is preferred for high-rate applications, and the 1-micron size allows for a high capacity at preferably low rates. The reason is the fact that the conversion of these powders to silver oxides during the charge leaves a core of silver metal of larger or smaller size at the center of the particle. When the finest powder, obtained from reduction of chemically prepared silver oxide, is used and reconverted electrochemically to silver oxide, the particles do not contain a core of silver metal and therefore electrodes fabricated with this powder will show a lower voltage and a very high capacity at low rates. As a matter of fact, cells using this powder exhibit a voltage dip when the discharge is started, and the voltage will recover when sufficient silver oxide is reduced to silver metal to compensate for the original lack of conductivity.

Another process is the use of silver oxide powder, which has a very fine particle size. In this case the powder is used as a slurry, which is pasted onto the grid with proper control for weight per square inch and proper thickness. The assembly is then sintered at high temperature to reduce silver oxide to silver powder.

The unformed silver electrodes are then assembled with unformed zinc oxide electrodes into a final cell, called dry unformed, which, when needed, is then filled with electrolyte and charged (a process called formation).

A variant is to charge the "green" silver electrodes after sintering in a separate tank, wash and dry them prior to assembly into a final cell, which is referred to as dry charged. When needed, the cell, after being filled with electrolyte, is ready for use.

Current Collector (or Grid)

Sometimes, a solid silver foil (usually 2-mil thick) is used. On each side of this foil is pressed a slab of silver powder of a controlled thickness, which is sintered under pressure and high temperature, until the desired designed thickness is reached. This method is useful when high conductivity of the plate is desired for high rate applications.

The solid silver foil is used when two slabs of silver powder can be made and then placed one on each side of the silver foil. The whole assembly is then hot-pressed, resulting in a sintering process of the silver powder slab and forging to the silver foil.

However, the most commonly used grid is made of an expanded silver metal (called Exmet from the company Exmet Corporation making it), which comes in various forms, thicknesses, and weights. To reach a relatively high conductivity, a dense pattern is sometimes used. It is usually not flexible or ductile, and therefore one should be careful to anneal the metal before use to avoid any spring effect, which has a tendency to curl the grid and the electrode at the edges. This has a detrimental effect because of the pressure exerted by the curled edges on the wet and weakened separator, which may result in a premature short at the edges. The final electrode should be flat before assembly to avoid this problem.

The Exmet grid should be centered inside the silver metal powder so as to have approximately the same amount of silver powder on both sides. In some cases, the grid comes out all in one side of the electrode, showing its pattern and bare metal. This type of assembly may cause damage to the cell. First there is the mechanical effect of pressure of sharp protuberances of the Exmet in direct contact with the separators. Also the solid metal of the grid facing the opposite zinc electrode will attract high current density, because of the highly sharp edges of the protuberances of the Exmet. The metal will be immediately converted to silver oxide during the first part of the charge, while the silver powder is still being formed, mainly on the opposite side of the electrode. This will create an imbalance during the formation between the two faces of the silver electrode and between the faces of the zinc electrodes opposite each face of the silver electrode, since current densities on each face of the silver electrode may be different.

The silver Exmet grid has to be of the proper pattern. Exmet Corporation uses a code related to the pattern giving the thickness of the stock silver material, the strand width, and the die configuration. From the mass m (grams) per in^2, one can calculate the equivalent thickness t (in mils) of the expanded metal of any type by dividing m by the density of silver (10.5 g/cc) converted to grams per in^3. The equivalent thickness t of an Exmet grid of a given area is the thickness of a solid silver foil of equivalent mass for the same area.

$$t = \frac{M \times 1000}{10.5 \times 16.4}$$

The equivalent thickness establishes the conductivity of the grid, for relative comparison, when selecting the type of Exmet grid to be used, depending on the desired voltage. A trade-off will be needed between voltage, weight, and cost. A low-rate cell does not need a heavy grid, while a high rate may require a high-conductivity grid.

It is worth noting here that the conductivity of the positive electrode, contrary to those in other batteries, increases as the discharge is proceeding, since silver oxides (which have a high electrical resistance) are converted to silver metal (which has a very low electrical resistance).

In all cases, whether solid foil or Exmet, the electrodes are cut sometimes from a large sintered silver powder sheet. The final electrodes must have round smooth edges to avoid sharp corners that may damage the separator, resulting in premature internal shorts.

Tab or Wires

The current collection to the terminal consists of several silver wires or a flat silver tab, whose cross section is sufficient to carry the current from each electrode, with acceptable voltage drop. The wires or the tab is hot-welded to the grid prior to the silver powder laying or pasting. The process can also be done after the plate has been sintered, since it consists only of pure silver.

A pull test is considered appropriate to ensure that the weld is acceptable. If spot-welding is done, several spots are required to ensure adequate welding for long life use.

Negative Electrode

The negative electrode active material consists of zinc oxide (or zinc powder, or electrodeposited sponge zinc metal) with some additives as deemed appropriate. The current collector or grid used is either a solid (or perforated) metal foil or an expanded or metal mesh grid.

Active Material

The most common electrode is made of zinc oxide of the

highest purity mixed with some additives (more detail on this later) pressed onto a current collector, with the use of some inert binder (such as Teflon powder or some CMC or PVA solution spray) to hold the zinc powder in place for handling purposes until it is assembled into a cell pack.

Sometimes, it is used as a slurry that is pasted onto the grid in quantity and thickness as required by the design.

In all these unformed electrodes, one finds it useful for easier production to use a layer of Aldex (tea bag paper, which is highly porous, thin, and light) on either side of the electrode to hold the zinc oxide for easier assembly into a cell. It may be wrapped around the electrode or simply placed on each side of the electrode. Sometimes it is removed prior to assembly or it is left, acting as an electrolyte reservoir around the zinc electrode.

The two types of electrode mentioned above can be used directly in cells intended to be shipped dry unformed, and which will be formed after being filled with electrolyte and charged. They can also be charged outside the cell in a separate tank. This process is referred to as electroformation; the electrodes are electroformed. They are then washed and dried prior to assembly into a final cell with fully formed silver oxide electrodes. The cell is then called dry-charged. When needed, after being filled with electrolyte, the cell is activated and ready for use.

Another process to obtain charged zinc electrodes is done by electrodeposition of sponge zinc metal onto a current collector.

This method is an electrolytic process. It makes use of an alkaline solution saturated with zinc oxide and of high-current density, so as to deposit rapidly a mossy or spongy form of zinc metal in sufficient quantity as required by design. The collector covered with sponge zinc metal is then removed from the bath, pressed sufficiently to hold the zinc, then quickly washed and dried to avoid rapid oxidation, which can lead to hot temperature and possibly ignition of the finely divided zinc. The final collector is then pressed to the desired thickness as required by design. It may then be cut to the proper size of the final electrode.

NOTE: The wet spongy zinc resulting from the electrolytic process must be washed completely free of alkali and dried properly to avoid instantaneous combustion in air and possible fire.

Additives to the Zinc Active Material

If zinc oxide is of the highest purity, it may not need any additives. Evidence of this has been found in some cells made with specially made zinc oxide. However, the smallest traces of contaminants, such as iron, nickel, chromium, can affect the self-discharge rate of the zinc electrode, thus leading to an unacceptable capacity loss.

It is therefore recommended to incorporate some additives in the zinc electrode active material to cover or minimize the effects of possible contaminants. The traditional additive is mercury in the form of an oxide or a salt, which after reduction leaves mercury in the form of a mercury-zinc amalgam. The purpose of the mercury is to cut on the gassing caused by the zinc in contact with the electrolyte and essentially reduce the self-discharge of the zinc electrode. The amount traditionally used varies from 0.5% to 2%, preferably 1% in most cases.

It has been found that gassing is reduced with increased mercury content, but levels off over 2% and no other benefit can be obtained over this value. On the contrary, it has been found that after cycling some actual mercury metal can deposit out the zinc electrode if the content is too high. With cycling, zinc metal is lost and detached from the electrode, and eventually too much mercury will be available in the remaining zinc.

Mercury has an advantage for primary cells (automatically activated primaries). It provides a better voltage regulation for these cells, which require the highest voltage for a short time.

Mercury has a disadvantage for long-cycling cells. It accelerates corrosion of the zinc electrode, often referred to as shape change, and eventually leads to shorter cycle life than cells not using mercury.

For this reason and for environmental reasons, mercury has been abandoned as an additive in all alkaline batteries using zinc electrodes. Several additives have been found acceptable (refer to U.S. Patent No. 4,084,047 of 11 April 1978). However, the most commonly used are lead and cadmium, sometimes each alone, sometimes in combination. The preferred combination, which has proved successful over many years in the field for long-life cells, is 2% cadmium and 1% lead. This is still currently and successfully used by the U.S. Navy. Lead oxide, being soluble in the electrolyte, eventually disappears from the zinc electrode. Analysis of cycled zinc electrodes has shown that lead disappears faster than cadmium from the remaining zinc-active material attached to the current collector.

Introduction of Additive into the Active Material

The most common method is thorough mixing of the additive in an oxide form with the zinc oxide power. The mixing of a very small percentage of additive oxides with a large quantity of zinc oxide is rather difficult and must be done in such a way as to result in a uniform material. Analytical tests must be conducted on various samples of the total mix, taken from different parts of the mix to ensure that there is no more than 0.1-percentage-point variation between samples.

When the zinc electrode is electrodeposited, the incorporation of additives can be done by a wet process, using for example acetate salts. The dry electrodeposited zinc electrode is dipped in a bath of known quantity of dissolved oxides and determining the time needed for a given pickup by the zinc plate. This is to be done experimentally, depending on the size of the entire zinc sheet and the size of the bath.

Current Collector (or Grid)

The negative collector must be made of a metal compatible with zinc and the electrolyte, leaving only the following metals: zinc, silver, copper, lead, and cadmium.

Zinc will eventually disappear as taking part of the oxidation-reduction process involved in the zinc electrode. It is sometimes being used as a partial addition to the zinc oxide material to strengthen the electrode. As a matter of fact, it could be used as active material in dry-charged cells. It comes under the form of a zinc foil, which could be folded as a U around the metal collector. A couple of cycles at a low rate are sufficient to get the cell to perform as usual, when the zinc metal (density 7.13) is actually converted to loose zinc, reaching a low density of approximately 2.5 after expanding in the space allowed by the proper cell design.

Silver is the metal of choice for most military applications where reliability and performance is of the utmost importance.

Copper is a possible substitute if plated with silver, lead, or cadmium. If the cell reverses inadvertently during discharge, copper will go in solution and will shorten the life of the cell. However, it is used to minimize cost of the cell, if cost is important.

Lead is another possibility if there is enough for conductivity purposes. Cadmium could be, but with the environmental pressure, it should be minimized. Both of these metals have a favorable effect on the zinc as additives as we have discussed above.

The configuration of the current collector is also important.

An expanded metal (like Exmet) or a mesh is sometimes used, but it is not recommended. The sharp edges protruding after the zinc electrode loses zinc at the edges are considered dangerous to the integrity of the separators pressing on the electrodes and may lead to rupture of the separators and eventually to internal shorts.

One practice is to use a solid perforated metal sheet with smooth edges and rounded corners, and of sufficient cross section to current carrying capability. If using solid sheet, the adhesion of the zinc oxide slab to the smooth silver foil will be enhanced by first cleaning the silver foil thoroughly with a solvent. Any grease or mold release material still on the foil may inhibit not only the adhesion but also the electrical contact and therefore the conversion of zinc oxide into zinc.

However the best practice is to use a perforated silver foil to hold the zinc-active material applied from both sides through the perforations, rather having a slab of zinc-active material on each side. A perforation percentage should be at least 40% so that both sides can participate in the electrochemical reaction in case that the quantities of active material are not perfectly balanced.

If the foil is too thin to sustain a large percentage of perforation, one should start with a thicker foil to make up for strength and for equivalent thickness as required for minimization of voltage drop. The perforation should be made before annealing the material for better results; the finished perforated sheet can then be annealed.

In order to get a smooth edge at the bottom and at the top of the collector, one could start with a strip of the electrode width (or slightly smaller) and start perforation about 1/8 or 1/4 inch away from the edges. The perforation should stop vertically when slightly less than the height of the electrode is reached so as to leave a solid piece between two successive perforation sets. It may be more expensive, but the benefit of not risking an internal short by the sharp edges of half-cut holes outweighs the possible cost increase.

Electrochemically, the highly perforated collector foil will balance the two sides of the zinc electrode and would not cause each side of the zinc electrode to work independently, so that sufficient zinc will be available for each electrode, even if the design is intended to provide excess zinc by calculation. Also, the end electrodes, sometimes fabricated as half-electrodes (or one-sided electrodes), need not be so. The excess zinc is always useful and it does not take up much space, considering the allowed expansion space left by design. Also, it provides better bonding of the zinc oxide slabs when there are two rather than one on a single side. Thus better quality assurance and less risk of mistake will be achieved by having only one type of electrode in production.

Tab or Wires

A flat tab or a few wires are connected to a top corner of the electrode to carry the current to the terminal. They should be of sufficient cross section to carry the current from the electrode. It is recommended to use silver metal for these leads, as they are in contact with the electrolyte and should be devoid of possible corrosion at the electrolyte-air interface. The leads are spot-welded to the grid in an area that has to be cleaned of active material. The weld has to be strong to avoid effects of corrosion, mainly if there are dissimilar metals involved. This is one more reason to use silver for the grid.

Separator

Cellophane remains the material of choice for silver-zinc batteries, in spite of efforts to introduce other materials. All commercial or military silver-zinc batteries use battery grade cellophane, which has been deglycerinated to the maximum. The thickness is 1 mil dry and it expands to about 3 mils when wetted by the electrolyte. It is used as is or treated with silver to enhance its strength.

The number of layers of cellophane separating the electrodes of opposite polarities varies, depending on the calendar life and the cycle life desired, on the rate capability desired, and essentially on the performance trade-offs desired.

Since the electrical resistance of each cellophane layer is approximately 10 milliohm-in^2 in 45% potassium hydroxide, it is obvious that a cell with 2 or 3 layers will be quite different in discharge voltage from a cell having up to 7 or 8 layers at the same current density. The three-layer cell is usually provided for short cycle and calendar life in some applications where no more than 3 to 5 cycles are needed over a period of 3 to 5 months. Such cells can then be designed for maximum energy density and relatively short soaking time (possibly one day). At the extreme range, a 7- or 8-layer cell is designed for long life, most of the time for long calendar life (possibly 2 years and over), but also generally for relatively low rate. However when it has to be filled with electrolyte, the soaking time may require up to 7 days to ensure that cellophane is fully wetted, mainly when the cell is tall. The cell may still work, even with some dry spots within the center of the cell pack, but because of the expansion of cellophane filling all the allowed space, these spots may never be wetted for a long time and may lead to some premature failure, due to some high resistance and uneven current density spots in these areas.

Other materials are used for providing a wet space at the face of each electrode. Usually a layer of nonwoven nylon (like one from Pellon Corporation) is placed on both faces of each positive and negative electrode, when room allows it. This layer, referred to as interseparator, comes in various thicknesses, from 2 to 5 mils and its selection depends on the size of the cell and its capability of being wetted in reasonable time, on the life expected, and on the energy density desired, each item being an element of the trade-offs.

Since the function of cellophane is to stop silver migration from the positive to the negative electrode and thus prevent the slow shorts, which eventually limit the silver-zinc cell life, a layer of another material placed in front of the cellophane layers or between the first two layers after the silver electrode has been found to be useful in delaying the silver migration and therefore the attack of cellophane.

One was found by chance when a layer of asbestos (10 mil thick, relatively iron-free) was used to provide a barrier against hot shorts (violent and fire-causing internal shorts) caused by accidental zinc dendrite penetration through the cellophane layers. This layer provides a retardant to the spread of heat and, being inorganic, could not sustain a fire between the electrodes. Since asbestos is being eliminated because of environmental concerns, a substitute is under consideration, which could be another inorganic material, like mats made of titanium oxide or zirconium oxide, which have the same properties as fire retardants (refractory) and compatibility with the electrolyte (KOH), but are relatively more expensive. The silver-stopping capability has to be demonstrated.

Another material that has been used several times is a film of PVA (polyvinyl alcohol), usually 1- or 1.5-mil thick and which does not expand in the electrolyte. It has a higher resistance than cellophane, but its silver-stopping capability is remarkable (refer to appendices in *Silver-Zinc Battery: Phenomena and Design Principles*, A. Himy).

Combining the good properties of both materials mentioned above, a layer of inorganic material, sprayed with or dipped in a solution of PVA, will provide a good fire-retardant and silver-stopper material to place between the first two layers of cellophane.

Cell-Pack Tightness

Allowance should be made for cell-pack tightness, considering in the design, not only the expansion of cellophane layers, but also the tolerances on zinc electrodes (usually 2 mils) and on silver electrodes (1 mil), mainly when the cell pack contains many electrodes. The resulting effect, if the pack is too tight when wet, is an undue pressure on the cell-case walls, thus causing a distortion of the cell, a tendency not to wet properly when filled with electrolyte (soaking), and eventually a prematurely poor performance of the cell.

Terminals

The terminals are usually hollow to allow for the passage of leads (wires or tabs) to be soldered. The terminals are made of copper silver-plated for high conductivity and corrosion resistance. For large cells, the tabs may be collected and mechanically attached to the base of the terminal, which does not need to be hollow.

Cell Case and Cover

The cell case and cover are made of plastic by injection molding. For easier release from the mold, there is always a taper from the top to the bottom of the case (larger width at the top than at the bottom), ranging from a few mils to more, depending on the height of the cell case.

The cover is cemented to the case with a solvent or epoxy, both resistant to potassium hydroxide.

Electrolyte

The preferred and traditional electrolyte is a solution of potassium hydroxide of 45% concentration.

This concentration allows the minimum degradation of cellophane, which is attacked as soon as in contact with the electrolyte, regardless whether the cell is cycled or not. The attack is more severe as the concentration is lowered, and at the extreme, water will attack and dissolve cellophane in less than three days at room temperature.

Although the concentration of 45% KOH allows for more zinc oxide dissolution (which is not desirable), it is still the lesser evil if one seeks long calendar life.

Some variations are often allowed for intended functions. For example, a lower concentration will allow a higher voltage, due to the fact that electrolyte resistivity decreases with lower concentration (minimum is reached around 31% KOH) as well as performance at low temperatures. For these reasons, the concentration is sometimes lowered to 40% and even lower in some special cases where cycle life is not as important.

The addition of zinc oxide to saturate the electrolyte in the hope that less zinc oxide will dissolve from the zinc electrode has been found useless. First, very little zinc oxide can be dissolved and it is very important that batteries from various manufacturers use a common electrolyte in the field, when needed, without resorting to a special electrolyte that has to be ordered or made under laboratory conditions. But more important is the fact that on recharge, after possibly only one cycle and certainly after two, all zinc oxide dissolved in the electrolyte replates onto the zinc electrode. If a cell is opened after a few cycles, some zinc oxide can be seen at the bottom of the cell, in part coming from shedding of the zinc oxide electrolyte, but also from excess not dissolved in the electrolyte.

Other additives to the electrolyte to limit the solubility of the zinc oxide during cycling are being tried. The important issue is not to lower the concentration of KOH to the point that it becomes detrimental to the cellophane.

An effect of cellophane when wetted with KOH is an absorption or reaction with KOH, which results in lowering the concentration of the surrounding KOH. If the cell is large or has many layers of cellophane, the drop in concentration may be very significant (from 45% to 38% in some cases). When there is time for replenishment of the cell because of loss of water by evaporation or by electrolytic action, the addition should never be made with water. It is to be remembered that water dissolves or degrades cellophane very fast, so that the top of the separators may be rapidly weakened. Therefore, the original concentration (45%) should be used, since very little would be needed and the original concentration has already dropped, as mentioned above. If frequent additions are required, the concentration may reach such a high level that formation of crystals may be detrimental to the separators. In general, the cell may not need frequent additions since it is never overcharged. As a compromise, a solution of approximately 35% KOH could be used for electrolyte complement, usually prior to the end of charge in order to have a thorough mixing of the electrolyte.

Formation

The process starts with filling the cell with electrolyte as experimentally determined by the manufacturer. The total amount may not go in at one single time. The process may require several attempts to fill the cell with the required amount, either with time (which may be too long) or by use of vacuum. The number of cellophane layers determines the soaking time, which may vary from one to seven days, to ensure proper wetting of all components. It is important to adhere to the proper soaking time recommended by the manufacturer on the basis of his experience for the type of cell delivered.

An unformed cell has to be formed by charging it at a low rate, not exceeding a current corresponding to 15 mA/in^2, possibly for 2 or 3 cycles if there are many layers of cellophane.

The discharge is to be done also at a relatively low rate, like the 8- or 10-hour rate.

A dry-charged cell is already formed and, after filling with electrolyte, is ready for use. It may be advisable to run the first discharge at a relatively low rate, possibly at the 3- or 5-hour rate.

Quality Assurance

Positive Electrode

The silver or silver oxide powder used in the manufacturing of the positive electrode should be of the highest purity. Which impurities and their maximum limits should be established by rigid specifications.

The particle size is important. The range found to be beneficial is 1 to 4 microns. However, the selection of the particle size of the silver powder is a design factor. A small particle size leads to a high silver utilization, i.e., a high capacity, but a relatively lower voltage at high rates. A large particle size leads to a relatively higher voltage at high rates but essentially to a lower capacity. This fact is due to the content of silver metal still left in the particle after it has been supposedly converted entirely to silver oxide. The voltage limit imposed for the end of charge does not allow for full conversion of silver metal to silver oxide. The particle size is therefore a factor in limiting this conversion. A large-size particle may still have enough silver in its core to allow for higher conductivity. A small-size particle may not have enough left, when the same amount of oxidation (or depth into the particle) is imposed on the particle.

The current collectors (grid and tabs or wires) should be made of pure silver metal. If an expanded metal is used, it should be annealed first, since the process of expanding the solid metal foil hardens the end product. The annealing will soften the metal so as to make the final electrode more flexible and less springy or curly, mainly if the expanded metal is heavy and of tight configuration.

The electrode should have rounded corners to avoid the possibility of damaging the separators by accidental piercing caused by sharp curled corners.

The thickness of the electrode should be uniform and maintained to the designed value within 1 mil tolerance.

Negative Electrode

The negative electrode is made essentially of zinc oxide and some additives. These materials should be of the highest purity, with impurity limits established by specifications. The most important impurities to control are iron, nickel, and chromium, which are conducive to local actions and self-discharge. It is desirable to have them completely absent from the materials.

The same purity standard applies to zinc metal, if used, either in powder form, or electroformed from zinc oxide, or electrodeposited from an electrolytic bath.

The thickness of a zinc oxide electrode cannot be held better than within 2-mil tolerance. The dry-charged electrode can be held to within 1-mil tolerance.

Separator

The absorbent material used as "interseparator" is usually porous and light and comes usually within 2-mil tolerance, but being easily squeezed and compressed, can be used with its nominal thickness for design purposes.

The cellophane, which comes as battery-grade cellophane, is treated to remove glycerin, which is detrimental to the operation of the silver electrode. The cellophane as procured should be inspected for resistivity in potassium hydroxide, thickness dry and wet (under a given pressure), and for possible defects as seen under microscopic analysis. A specification should be established for inspection.

Electrolyte

The potassium hydroxide should have the highest purity and be free of possible contaminants. A very strict specification should be established for procurement. Carbonate content should be less than 0.5% when received.

CHAPTER III

Characteristics and Operation

The following table is an abbreviated list of the principal characteristics of the silver-zinc battery:

Open circuit voltage (OCV):	1.86 V at room temperature
Closed circuit voltage (CCV) or operating voltage at the C/5 rate:	1.55 V
End of discharge voltage:	1.2 or 1.3 V at the C/5 rate
(The capacity left after the end of discharge voltage is very small)	
Self-discharge per month at 20°C:	5%
Capacity change from 20°C	
to 0°C:	5 to 10% depending on the rate.
to 40°C:	0
Operating temperature range:	-20°C to 60°C
Storage temperature range:	
Dry unformed:	-40°C to 60°C
Dry charged:	same
Wet discharged:	-30°C to 50°C
Wet charged:	same

(avoid temperatures higher than 40°C for long periods)

Preferred storage conditions, when wet (charged or discharged):	5 to 10°C

Life capability:
(in protected conditions, cells sealed)

Dry unformed:	25+ years
Dry charged:	up to 10 years
Wet discharged:	up to 2 years (room temperature)

Operational calendar life (wet, sometimes standing charged and some cycling):	2 years at room temperature

Cycle life (100% depth of discharge, room temperature, C/5 rate) (for cells designed for long life):

down to 80% original capacity:	40 to 50 cycles
down to 60% original capacity:	100 cycles
end of life (shorting):	100+ cycles

Capacity loss:
20% after 50 cycles
40% after 100 cycles

(However, the original capacity is maintained for several cycles before dropping gradually to approximately 80%.)

Energy density: (at the C/5 rate):

	gravimetric	volumetric
Cell basis:	140–180 Wh/kg	400–500 Wh/L
Battery basis:	100–160 Wh/kg	200–450 Wh/L

Power density: (Short Duration Pulses)

Fully charged to half-discharged:	500 W/kg
If specially designed:	up to 2200 W/kg

Gassing (in cc/hr per rated 100 Ah capacity, at 20°C):

On stand:	approximately 2 cc
On charge:	0 cc up to 95% of the charge 2 cc for remainder of charge up to 2.0 V

Method of charge (preferred):

Constant current to first cell to reach voltage cut-off:	2.03–2.05 V per cell for vented cells 1.97–2.00 V per cell for sealed cells

Recharge time:

Low-rate cells (high energy density):	up to 72 hours
High-rate cells:	8 hours

Operation

The success of the operation of a silver-zinc battery is its reliability. Reliability is defined as the probability that an item can perform its intended function for a specified time under stated conditions.

Therefore it involves: intended function, design, manufacturing, reproducibility, maintainability, and operation.

Intended Function

The key to reliability is to have an overdesigned battery. It is a fact that the silver-zinc battery deteriorates with age, not only with cycling and usage. It is also known that a battery of several cells would give fewer cycles than a single cell; the greater the number of cells, the fewer the number of cycles.

Therefore, the capacities of the cells to be used to form a battery should be matched within a narrow range. It is not uncommon to match them preferably within 3 sigmas. For this reason, the cells should be fabricated unformed, so that the formation cycles can give the user an opportunity to reject the cells falling outside the specified capacity range. A dry-charge battery would not take advantage of the selection process.

Design and Manufacturing

A few repetitions are summarized here.

The positive electrode is to be made of pure silver powder with a particle size preferably under 3 microns. The grid is to be made of pure silver metal (usually expanded metal or solid or perforated foil). If the electrode is relatively thick, the grid should be centered. The electrode thickness should be within no more than 1 mil.

The negative electrode is to be made of pure zinc oxide powder with additives such as lead and cadmium (in lieu of mercury). The grid is to be made of pure silver (expanded metal or perforated foil) again centered within the powder. The powder is usually held with a binder and covered in a porous paper bag for easier handling. The tolerance on the thickness of the overall electrode should not exceed 2 mils.

The mass ratio of zinc to silver is in the range of 0.9 to 1.2, depending on the acceptable capacity loss over cycle life, compatible with the maximum energy density. There is no need to have excess zinc over silver since this will detract from the energy density without significant increase in cycle life.

The separator system makes use of a so-called interseparator, which is an absorbent material placed on the face of the electrodes such as a nonwoven polyamide (nylon), preferred to a woven nylon that does not retain electrolyte as well. The main separator is for the time being exclusively cellophane (1-mil-thick dry). A silver stopper, such as a polyvinyl alcohol (PVA) film, is sometimes used in front of the main separator, between the interseparator and the main separator, for the purpose of retarding silver penetration. The wrapping is usually done by bagging a pair of positive electrodes, so that the negative electrodes lie outside the positive bags.

The electrolyte consists preferably of a solution of 45% KOH. Any additives are still in the process of investigation.

The number of electrodes and the surface area of each electrode should be designed in such a way as to accept no more than 3 mA/cm^2 for charge current.

Maintainability

Maintainability is the measure of the ability of a battery to be retained in (or restored to) a specified condition when maintenance is performed according to prescribed procedures. For silver-zinc batteries, the maintenance consists simply in maintaining the electrolyte level. Since very little gassing occurs, the addition of electrolyte is infrequent. It is preferable to add only a dilute solution of electrolyte, and not water, since water staying at the top degrades cellophane faster. Since a dilute solution of KOH may not be easily available, a standard electrolyte (45% KOH) is used instead.

When not in use, the battery should be stored in a discharged state to avoid silver attack on the cellophane and so prolong battery life. If charged, the battery should be stored in an ambient temperature of approximately 5° C.

Operation

The charge voltage should be limited to less than 2.05 V per cell on an individual cell basis, preferably 2.02 V and sometimes 2.00 V when the battery is on float. A voltage scanner is sometimes used to ascertain that all cell voltages are scanned and that the charge is discontinued when the first cell reaches the specified cut-off voltage.

The discharge should be limited to 1.3 V per cell. Practically very little capacity is left beyond the 1.3 V limit. When possible, the limit could be set at 1.4 V, which would prevent the possibility of reversal when a large number of cells are in series. Also, if the cell is overdesigned, as is usually done, there is no need to remove more than 100% of the rated capacity, which is what is required for the usage.

Life

If we define the life of the battery as the point at which the battery is no longer able to perform its rated function, the experience obtained in some batteries in the field is as follows:

If the battery is most of the time in a float condition with few cycles usage, its life can be reliably 2 to 2.5 years.

If the battery is used at 50 to 80% depth of discharge several times up to 50 cycles with periods of charged stand in between, its life is usually in the range of 1.5 years.

This is documented by experience for batteries used under field conditions with a varying temperature range. Single cells cycled under controlled laboratory conditions can give up to 100 cycles at 100% depth of discharge.

CHAPTER IV
Discussion on Various Subjects

This chapter covers various issues that are often raised when dealing with silver-zinc batteries as well as some phenomena.

Filling and Soaking

The process of filling a silver-zinc cell may be laborious due to the fact that cellophane does not wet and absorb the electrolyte as readily as one may wish. The number of positive electrodes determines the total number of cellophane layers, since each face of electrode is covered with so many layers of separators. If n is the number of electrodes and s is the number of cellophane layers, the total number of layers will be F = 2 ns. This number determines the soaking time T (in hours) needed for a given silver-zinc cell. It may vary from one hour to a few hours for a very thin separator system and a few electrodes, to 7 days for a thick separator system and a large number of electrodes, and very often with the help of vacuum to make the cell accept the required amount of electrolyte. An empirical formula gives the soaking time T as a function of the total number of cellophane layers:

$$T(hrs) = 0.54F - 3.3$$

It is close enough as a first approximation, and the result can be rounded off to the next hour or if over twenty-four hours and converted in days to the closest integer.

The manufacturer has previously determined how much electrolyte the cell requires for optimum performance, considering that the final level of electrolyte should not exceed the top of the separator during operation. Calculation of the amount of electrolyte can be made by finding the free volume left in the cell case, after considering the porosity of all electrodes and porous absorbent materials, and taking into account the swelling of cellophane layers. This amount is then verified experimentally by testing the cell during formation, which usually requires 1 to 3 cycles, depending on the size of the cell.

A fact little known is the presoaking of cellophane in potassium hydroxide. If one lets cellophane soak in KOH, then dries it (by removing the water content), the end result is a compound of potassium ion combined with cellulose, making a readily wettable membrane as soon as in contact with water or in the case of a battery with a solution of potassium hydroxide. This may be useful in some manually activated primary silver-zinc batteries in which fast activaton is desirable, while only 2 or 3 cycles are required. It still remains to be fully tested in secondary cells if there is some specific application, mainly to ascertain if calendar life and cycle life are affected.

Charge

The charging of a silver-zinc cell is very sensitive to the cut-off voltage determining the end of charge. First it is established by the optimum current density applied to the electrodes. By experience, it has been found optimum not to exceed 20 mA/in^2 applied to the surface area of the total number of positive electrodes.

Since the positive electrodes control the capacity of the cell, their number is usually even and the number of negative electrodes is odd. Here is an example of calculation for determining the charging current. If the electrode area is 9 in^2 and the total number of positive electrodes is 8, the total area to be considered (counting the two faces of each electrode) is

$$2 \times 9 \times 8 = 144 in^2$$

The charging current then cannot exceed

$$144 \times 20 = 2880mA$$

which, rounded off, can be set at 2.8 A or lower.

A lower-charging current would result in a better electrochemical conversion and therefore better efficiency.

The cut-off voltage then is a function of the charge current. With a high-charge current, limited as shown above, the cut-off voltage can be set at 2.05 V per cell. With a lower charge current, it can be set as low as 2.0 V per cell.

Ideally, a current density of 15 mA/in^2 corresponds to a cut-off voltage of 2.02 V per cell under controlled conditions. This means that a battery consisting of several cells in series should not be charged if any cell exceeds 2.05 V at any time. Usually a voltage scanner connected to the terminals of all cells in the battery signals if a cell reaches or exceeds this value.

The overcharge of a silver-zinc cell, over the value of 2.05 V, results in zinc penetration, which eventually contributes to the shorting of the cell, resulting in premature failure of the battery. A cell that is never overcharged or never sees an end voltage over 2.05 V will never fail by zinc penetration, only by silver penetration through the several layers of cellophane, eventually causing a slow short and ultimate failure of the cell.

Several batteries making use of scanners have been used in such a manner in the field and have lasted two years, with low capacity considered as failure criterion, rather than shorting.

Electrolyte Level

The electrolyte level is indicative of several factors in the operation of the cell.

When the design makes use of a so-called "normal" wrap,* where each set of two positive electrodes are wrapped in several

*A. Himy, *Silver-Zinc Battery: Phenomena and Design Principles,* (New York: Vantage, 1986).

layers of cellophane as a U and the negative electrodes are standing outside the U, the end of charge is characterized by a high level of electrolyte, which should be adjusted not to exceeed the top of the separators. This is due to the fact that on charge, the hydroxil ions (forming water molecules) are moving toward the negative electrodes, while on discharge the reverse occurs. If gassing occurs toward the end of charge (because of overcharge) or even during stand (because of self-discharge of the zinc electrodes), the gas bubbles trapped in the electrolyte between the negative electrodes and the separators contribute to the high level of the electrolyte. It is therefore necessary to remove them by applying a slight vacuum to the cell so that the level of electrolyte is more representative and can be adjusted accordingly, always at the point where the charge is considered terminated.

Obviously, if the design of the cell pack makes use of a "reverse" wrap (prior reference), where the negative electrodes are wrapped, leaving the positive electrodes outside the U's, the reverse happens.

This design is not frequently used, because the accumulation of zinc (or zinc hydroxide) due to shedding or swelling of the zinc electrode at the bottom may cause a rupture of the bag and thus lead to premature failure.

In either case, after the end of charge and after the cell is left standing for a while, the level of electrolyte will drop to a medium level, when equilibrium between the levels inside and outside the U separator bags is reached. If the cell is left standing charged over a long period of time (several weeks), the level may happen to drop even lower, mainly when the cell has a tall configuration. Caution should be exercised not to add electrolyte at that point. The cell must first receive some booster charge as appropriate.

Long-Term Charged Stand

When a charged cell has been standing for a long time (several weeks or months), and one wants to give it a booster charge with the normal charging rate, the voltage jumps quickly to the cut-off point recommended not to exceed. A high-resistance layer may have been formed on the surface of the electrodes, probably the zinc electrodes, either by carbonation or by drying. Therefore it will be advisable to start with a lower-charging current until this layer is broken, or the electrolyte starts rising.

It has been reported sometimes that the open-circuit has dropped somewhat from the original 1.86 V usually considered as indicative of a fully charged fresh cell. This phenomenon may be caused by the formation of some silver monoxide, which gives rise to a mixed potential. Of course, if the cell is old, it may be the sign of an incipient internal short, caused by the slow migration of silver through the separators with time. It is to be remembered that silver oxides in potassium hydroxide solution will deposit as silver in the cellophane, layer by layer, until a trail is formed and comes in contact with the zinc electrode, thus causing an internal short.

There may be an operating condition where the cell is continuously connected to a power source so as to be on charge float for a long time when not in use. In this case, the electrolyte level will be always high (if the cell design uses the "normal" separator wrap). This condition is conducive to a possible drying of the surface of the positive electrodes, mainly when they are of a tall configuration. It is therefore imperative to design the cell-separator system with a thick absorbent material (nonwoven nylon or Pellon) placed against the positive plates so that some electrolyte can be retained by absorption during the stage of electrolyte motion toward the negative electrodes. It has been frequently found that failed cells operating in this condition exhibit upon dissection an accumulation of crystals on the surface of the positive electrodes, a sign of water deprivation. The crystals may be a form of potassium oxalate, resulting from the deposit of concentrated potassium hydroxide reacting with the organic material of cellophane. The presence of these crystals may be harmful; when under the internal cell-pack pressure during cycling conditions, the crystals exert pressure points against the cellophane, thus causing punctures leading to internal shorts.

One way to avoid this drying effect caused by long-term float charge is to discharge the cell periodically, using some shallow discharge only.

Life

For a silver-zinc cell, one should consider two types of life: cycle life and calendar life. A cell may be tested on cycling, under a fast, automatic-cycling mode, resulting in a relatively large number of cycles. But this may not be representative of the actual usage of the battery in the field, which may require periods of charged stand over a long time, with periodic usage of the battery with various depths of discharge. This is more representative of calendar life.

A battery that has been wetted and formed and cycled 2 or 3 times for verification and which may be left on charged stand over a long period may not have the proper calendar life, even if not used at all during that period. Silver migration (or penetration through the separators) and constant cellophane degradation in the electrolyte will shorten the battery life. It is therefore important to design the separator system for that specific application if long life is needed. The number of cellophane layers will determine the calendar life. However, this approach will detract from high rates and from high-energy density. A trade-off has to be established. The high rate can be properly taken care of, by adjusting the total electrode area so that the desired voltage can be obtained, but the energy density will certainly suffer.

It has been found that a maximum of 8 cellophane layers is sufficient for long life, considering that cellophane cannot last more than 3 years in potassium hydroxide at room temperature. Therefore, increasing the number of layers is useless for life and will detract from the energy density. As explained previously, a silver stopper or barrier can help reduce the rate of silver migration and attack on cellophane, thus making possible the advantage of a longer calendar life.

Electrochemical versus Chemical Silver Oxides

A little-known phenomenon is the open-circuit voltage exhibited by a silver-zinc cell where the positive electrode is using silver dioxide made chemically rather than electrochemically. The silver dioxide, chemically made, can be pressed onto a silver grid to fabricate a readily charged positive electrode. The open-circuit voltage then exhibited is not what will be expected for an electrochemically charged silver electrode where the silver has been electrochemically converted to silver dioxide. Whereas such a voltage (of a freshly charged cell) is known to be 1.86 V, the open-circuit voltage of a charged cell using chemically made silver dioxide is only 1.65 V. When the cell is discharged and then recharged, the open-circuit voltage reverts to the usual 1.86 V. It appears that the corrosion of the silver grid (conversion to silver oxides) during the recharging process is the cause of this voltage-value change from chemical (where there is simple contact) to electrochemical (where the bond is more than simple contact).

This phenomenon is useful in the fabrication of button cells where the voltage is desirable to be low (1.65 V) while the capacity is desirable to be high (by using silver dioxide). It is also used in some primary cells where there is a requirement for maximum voltage regulation, where the electronics cannot accommodate a too high open-circuit voltage (see the following paragraph for other means to lower the open-circuit voltage).

Suppression of Silver Dioxide Open-Circuit Voltage

If a limit on the open-circuit voltage is desired, only silver monoxide can be used by pressing or pasting such an active material onto the metal grid, of course only in the case where one single discharge is required. However, this method will yield only half the capacity one could otherwise get from silver dioxide. If more capacity is desired as well, then one means is to predischarge the fully charged silver plates outside the cell to a level where the potential is below the maximum open-circuit voltage. But in this case, one may obtain so-called mixed potentials between the maximum value of 1.86 V and the value of 1.65 V. Even so, the voltage obtained may not be stable and a fine experimental tuning is needed to determine exactly the predischarge output so as to obtain the maximum open-circuit voltage desired. This method is therefore not recommended.

Better means have been devised to lower the open-circuit voltage of a fully charged silver plate from the 1.86 V to the 1.65 V level in the case of primary batteries.

Heat Treatment

The fully charged plates can be immersed in a pure hot water bath for a certain time to convert a thin layer of silver dioxide to a silver monoxide layer, thus getting a lower open-circuit voltage reading during the short electrolyte filling time required to activate the cell. The bath should contain pure distilled water maintained close to 100° C and the time of immersion depends on the number of plates immersed and their size and is determined by experience, but is in the order of 10 minutes to 1 hour. A sample plate could be removed at various times and its voltage could be measured against a reference electrode made of zinc foil or wire.

Chemical Treatment

The fully charged electrode (dry) is dipped into a very dilute solution of silver nitrate, then dried. When in contact with a solution of potassium hydroxide, there occurs a conversion of some silver dioxide to silver monoxide, which then depresses the open-circuit voltage. The potassium hydroxide treatment can be done before using the positive electrode into a final cell or else the plate can be used as is before the potassium hydroxide treatment, so that the conversion occurs during the filling (or activation) of the cell with potassium hydroxide.

Reversal of the Cell

The silver-zinc cell is not amenable to reversal condition in the case of overdischarge, when a cell of low capacity is driven into reversal by other cells in series. Although the cell can be recovered, its life may be eventually shortened. At first zinc in solution will plate onto the silver electrode, but eventually upon recharge, this plating will go back into solution. Zinc is also forced within the pores of the separators, being eventually conducive to the easy formation of dendrites in subsequent cycles.

It has been also established that reversal is not the only condition where dissolved zinc plates onto the silver electrode. It is sufficient to short the terminals of the cell (as frequently done for nickel-cadmium cells if one desires to fully discharge the cell), to have caused zinc plating onto the silver electrode. As a matter of fact, even discharging below 0.5 V will cause such plating. This is why it is recommended not to discharge the cell below this voltage when the cell has to be fully discharged for handling for shipment.

Current Collectors or Grids

It has been sometimes reported that metal other than silver could be used as current collector or grid or even tabs in the positive electrodes. Although the cell works to a certain extent, long life and good performance of the cell are still not optimum. Nickel has been used, but it depresses the voltage of the silver electrode, causing inefficient charge, with gassing before the desired end of charge. Copper is obviously out of question, since its oxidation and subsequent dissolution of copper oxide will harm the separators; even silver-plated copper may not be sufficiently protected, as the silver plating is too thin and gets oxidized very rapidly so that copper may be leached out.

In the negative electrode, plated copper may be less harmful, except in the case of possible reversal of the cell. Plating can be done with silver, lead or cadmium. Even the entire grid could be made of lead or cadmium, if the cell is not operating at high rates. However, if cost is not an issue, only silver metal should be used as the preferred material for optimum performance and reliable life.

When copper is used without plating, it should be remembered that in the presence of mercury as an additive in the zinc-active material, only one cycle is required for the copper to be entirely plated with zinc metal. Although some gassing may be observed in the beginning in the first cycle, the gassing due to copper in contact with zinc is practically insignificant immediately after the second cycle.

Internal Shorts

The silver-zinc cell system has a peculiar feature different from other battery systems with respect to the open-circuit voltage (OCV), which gives an indication of the possibility of an internal short. When the cell is freshly charged, its OCV reads 1.86 V at room temperature and is stable over a long period of time. In other battery systems, the OCV is not fully defined and there is always a decline due to a slight self-discharge or instability of the active materials. In the silver-zinc system, the silver dioxide is relatively stable over several days at room temperature so that any decline from the OCV value is attributable only to an internal short. When one may see a decline of a few millivolts (from 1.86 to 1.85 V, then 1.84 V) over a period of one or two days, one can conclude that a slow internal short is present. This is usually the case when the cell has reached a cycle life or calendar life where silver migration through the separator layers has reached the last layer and a silver path then exists between the opposite electrodes. This short is not enough to create a violent short resulting in overheating and eventually a fire. But it is an indication that the end of life is imminent. A short due to zinc penetration is usually more severe, in that the drop in OCV value is faster, depending on the importance of the zinc dendrite presence. A real severe zinc penetration (by zinc dendrite) results in rapid overheating and possibly fire, if the cell has a large-energy storage capacity.

One test that should be done periodically to determine the health of the cell is to give the cell a fresh charge, let it stand overnight or a couple of days, and periodically measure the open-circuit voltage. The voltage stability (or instability) will give the user the option of continuing the operation of the cell in a possibly critical mission, mainly when the cell has reached a point in life where there is some doubt.

Zincate in Electrolyte

When an unformed cell is first filled with the potassium hydroxide solution electrolyte or when a cell is being discharged, the electrolyte dissolves some zinc oxide (or hydroxide) up to a certain amount determined by the electrolyte concentration, the temperature, and the availability of electrolyte. The most important factor is the amount of electrolyte available around and on the top of the cell pack, so-called "free" electrolyte since the rest is absorbed in the pores of the electrodes and the separators. Therefore, the total amount of zinc oxide dissolved is minimal. Upon recharge, the zinc plates out of the solution onto the zinc electrode, practically depleting the electrolyte of zincate. The continuation of charge into an overcharge mode creates gas evolution on the zinc electrodes (hydrogen) and contributes to the formation of zinc dendrites. It is therefore important to limit the charge to a voltage below the gassing potential of the zinc electrode (approximately 2.1 V). A practical limit is always 2.05 V for a single cell and even less when the battery consists of several cells in series. For example: 2.03 V × number of cells, when the battery consists of a large number of cells. It is considered better to have the battery slightly undercharged than to expose it to an accidental overcharge of one or more cells. In such a case, when feasible, a voltage scanner is connected to each of the battery to determine the cut-off point of the charge, when a cell reaches 2.05 V, in which case the charge is completely cut off or the charge current is lowered.

A misconceived notion to avoid dissolution of zinc oxide into the electrolyte is to fill the cell with an electrolyte saturated with zincate so as to avoid loss of zinc oxide from the unformed zinc electrodes. First it has been established through several experiments and analyses that the electrolyte is immediately filled with zincate after one single cycle. On the other hand, the amount of zinc oxide dissolved from the zinc electrodes is minimal. More zinc oxide is mechanically removed by shedding from the electrodes either in the dry state or first in contact with the electrolyte and deposits at the bottom of the cell around the cell pack. This state is not harmful to the cell operation, although it is not an appealing sight. In any case, the zincate in solution disappears immediately on recharge since it is plated out onto the zinc electrodes at the end of charge.

One must also remember that the zinc oxide electrodes are always designed with excess active material so that a small loss of zinc oxide either mechanically or by dissolution is not going to impair the performance of the cell. The cell failure is due to some other causes than to a small loss of zinc oxide at the beginning of the cell life.

NOTE: The same holds true of other rechargeable alkaline zinc systems, such as nickel zinc and zinc air.

When the zinc oxide is dissolved in the electrolyte, it is not restricted to the solution around the negative electrodes, but diffuses through the separators to reach the silver electrodes. It is not harmful to the operation of the cell, but it may sometimes be found in white deposits (zinc hydroxide) in areas between the silver plates and the separators, even possibly in the pores of the silver electrodes, but without any noticeable effect. It has been reported that the effect is more pronounced in the nickel electrodes where blockage of the nickel-active material can affect the active material discharge after several cycles. In the silver-zinc cell, the effect may not been noticed since the silver-zinc cell does not reach a large number of cycles and fails for other reasons. It may be worth investigating this phenomenon, since in some cases, a permanent capacity loss is experienced without being traced to either electrode or to the separators, as when the cell had a few cycles and is left on charged stand.

Effect of High Temperature on Cell Life

When a cell is filled with electrolyte, high temperatures due to simple exposure or too high discharge rates are conducive to fast degradation of the separator (cellophane) by the electrolyte. This in turn will reduce the cell-cycle life and calendar life well below what is expected from a cell performing at room temperature. High temperature is defined here as over 55° Celsius. For example a cell designed for an 18-month life at room temperature may last only approximately 9 months if high-rate discharge is performed at every cycle ending with a temperature higher than 90° Celsius. Such a cell may have been designed with 6 layers of cellophane for operation at room temperature or at least not higher than 56° Celsius. But at the high temperature mentioned in the above example, the 6-layer separator system may be equivalent to no more than 4 layers.

Another effect of high temperature also contributing to the cellophane degradation is the acceleration of the silver penetration through the separator layers when the cell is left on charged stand for long periods. The silver attacks the cellophane even more at high temperatures.

Capacity Loss

The capacity loss of the silver-zinc cell is attributable to several factors, depending on the conditions involved.

During continuous cycling, the capacity loss is entirely due to the zinc electrode shape change, loss of material, densification at the bottom of the electrode; very little is attributable to the silver electrode.

During charged-stand storage, where the cell is left standing fully charged and unused for long periods, the capacity loss may be due to decomposition of silver dioxide to silver monoxide, thus losing up to 30% if the cell is left on storage for very long periods. Otherwise, the capacity loss may not be more than 5% per month. The open-circuit voltage (OCV) may be a good indication after long periods, but not in the beginning (if a decrease in OCV is detected in the beginning in a matter of days, it is a sign of an incipient short). The self-discharge is due entirely to the silver electrode, not to the zinc electrode, which is designed with excess zinc capacity. The zinc electrode self-discharge is minimized by additives to the active material. It still does not account for the capacity loss found after long storage times. One may claim that the zinc gets "passivated," but a discharge performed immediately after a long stand does not show capacity limitation by zinc.

It has been reported sometimes that this capacity loss is permanent and cannot be recovered. When the cell is left standing for a long time, all electrolyte is drained to the bottom of the cell and the top may be relatively dry or at least devoid of sufficient electrolyte to carry the current in the upper part of the cell pack. It may then take some time to raise the level to the proper height, which can be done by using a low-rate charge for a relatively longer time than usual. The silver electrode may be partially discharged, and the zinc electrode may be close to full charge. This imbalance may be corrected carefully by using this low-rate charge. This procedure may have to be done for a couple of cycles.

A simple method for determining the cause of the capacity loss is by opening the cell in a discharged state and examining the areas of a silver electrode and its opposite zinc electrodes. If the zinc has some shape change, it will be mirrored in the bottom white (discharged) portion of the silver electrode facing the remaining zinc (or zinc hydroxide) still attached to the grid. The top portion will be still black. If the cell is opened in a charged state, one may be able to see that the bottom of the silver electrode is very black (fully charged) while the top may be either white (discharged) or grayish (partially charged). In this case, the cause of the capacity loss is obviously due to the absence of zinc at the top of the zinc electrode. During the charge, the lines of current between the top portion of the zinc electrode grid to the top portion of the silver are not as uniform as those between the bottom portions.

If there is no shape change, then the silver electrode is entirely black in the charged state or all white in the discharged state and the cause may be found in the silver electrode, due to the decomposition of silver dioxide to silver monoxide. A verification can be done by recharging this electrode separately against dummy electrodes and determining its real capacity, which can be found to be very close to the original capacity of the electrode (this can be determined by weighing the electrode washed and dried and subtracting the weight of the included grid and computing the silver utilization).

Capacity loss due to the decomposition of silver oxides is usually very small. (Salkind discusses this in more detail in "Self-Decomposition Processes in Silver Electrodes." See bibliography.)

Crystals in Aged Cells

Sometimes upon dissection of aged, used or failed cells, one finds some crystals of various sizes between electrodes and separators. These crystals, if large enough, are capable of applying pressure against the separator layers, weakening their strength, and eventually piercing them, thus leading to an internal short. These crystals have been sometimes identified as potassium oxalate resulting from decomposition (or hydrolysis) of cellophane in potassium hydroxide medium. They may be also at times only potassium hydroxide crystals or potassium carbonate crystals deposited from a highly concentrated potassium hydroxide solution. They are usually found on the sides of the positive electrodes, which, in the most common and traditional design, are separately bagged in a separator "U." During charge, water moves from the positive electrode bags to the outside toward the negative electrodes so that the level of electrolyte rises until it reaches a maximum at the end of charge. During this process, the remaining electrolyte in the positive bags becomes concentrated more and more so that given certain conditions, crystals can form, precipitated out of the solution. During discharge, water returns to the positive bags but may not be able to dissolve the crystals if they are large enough or if there is not enough water, mainly when the cell has been overused or has seen prolonged charging, usually on float. The location of the crystals could either be close to the positive electrodes or between the internal cellophane layers inside the positive bags.

CHAPTER V
Zinc Electrode

This important subject requires a separate treatment by itself, since at the present, the zinc electrode is the limiting factor in all alkaline zinc batteries, such as silver-zinc, nickel-zinc and zinc-air. These batteries have many common features and many common problems. (A comprehensive summary on the zinc electrode investigation can be found in McLarnon and Cairns and in Chen et al; see bibliography.)

Shape Change

It is a well-known phenomenon that the zinc electrode loses zinc-active material during cycling of the cell. Most remarkably, the shape of the electrode changes faster at the top and on the upper sides of the electrode, while the bottom of the electrode grows thicker. During charge, the zinc in solution does not replate onto the electrode at the same sites; it replates preferentially at the bottom where the electrolyte is richer in dissolved zinc hydroxide. Still, this may not be a complete explanation.

The traditional additive to the zinc active material had been mercury (generally in the form of mercuric oxide) to form an amalgam with zinc metal. This additive causes a faster shape change than in cells containing mercury-free zinc or zinc in which other additives have been incorporated (such as lead, cadmium or calcium).

With respect to the contribution of mercury to shape change, a study has been made with a radioactive tracer technique (Einerhand et al., "Zinc Electrode Shape Change," *ECS Journal*,

January 1991, pp. 1–17). A summary of the findings is presented in the following.

A radioactive mercury was incorporated into the zinc electrode to monitor the spatial distribution of the radiotracer over the electrode during repeated cycling. After several cycles, the cell was disassembled and the mass of zinc and the radioactivity of each region of the zinc electrode were measured. They show the same pattern as the shape change, little at the top and most at the bottom.

Cells containing mercury lose capacity during cycling faster than cells containing no mercury, and their cycle life is shorter. It was observed that the radiotracer distribution over the electrode does not change significantly. The paper concludes that "zinc material transport as a result of detachment of electrode mass is highly unlikely since, if this type of transport occurs, transport of mercury must be observed."

The paper also emphasizes that "during cycling, the mercury-zinc ratio over the electrode varies considerably since the zinc material distribution changes and the spatial distribution of mercury does not change. The mercury-zinc ratio decreases at (locations) where zinc material accumulates (bottom) and it increases at (locations) depleted in zinc material (top)." The authors remark that "if processes at the zinc electrode are affected by the mercury additive, the differences in mercury-zinc ratio over the electrode enlarge its effect."

NOTE: This means that zinc at the bottom of the electrode during cycling becomes less amalgamated and zinc at the top becomes more amalgamated, thus creating regions of dissimilar alloy content in the zinc electrode, which may start parasitic local actions. This, in turn, contributes more to self-discharge, corrosion effects, and therefore shape change.

In the case of mercury-free additives, these additives disappear at the same rate as the zinc during cycling (assuming that they are initially uniformly mixed with the original zinc oxide active material). This would not create a corrosion effect since the remaining zinc at any time during cycling will still be uniformly mixed with the additives in the same ratio.

Dry Charge

It has been found convenient in some applications to have the cell dry-charged, meaning that the electrodes have been assembled in a charged state, having been formed separately and dried prior to assembly.

The silver electrodes are converted to silver oxide by an electrolytic process in a potassium hydroxide solution (not necessarily full strength), then washed free of KOH and dried. The electrode is then ready for assembly. Some manufacturers paste the grid with a slurry of silver oxide; the resulting electrode is then pressed to the desired density, then dried. This process does not lend itself easily to a controlled mass and thickness of the resulting electrode.

The dry-charge process for the zinc electrode is more involved. In chapter 2, under "Negative Electrode," it was reported that two processes are in use: electroformation and electrodeposition. The electroformation consists essentially of pasting a zinc oxide slurry onto a grid and forming a green plate, which is then converted to zinc in an electrolytic process. The electrodeposition process consists in depositing zinc metal sponge onto a collector from an alkaline zincate solution at high-current density so as to obtain a spongy zinc. In both cases, the resulting electrode is washed free of KOH and then dried to avoid oxidation in air.

Usually, a small amount of zinc is oxidized, but it does not impair the performance of the electrode, since it is designed with excess zinc.

However, an important point is to be made here. Because the zinc electrode is supposed to be fully charged, the dry-charge cell can be discharged immediately after filling and an appropriate soaking time, which is a relative advantage over the unformed cell. The only difference is the time saving of formation of the unformed cell, which may require two cycles. If the dry-charge cell is to be checked before the actual application usage, then the difference is only one cycle. Very little time indeed, if we consider one specific disadvantage. Upon recharge and at the end of charge, the dry-charge cell reaches a state in which the silver electrode is fully charged, but so is the zinc electrode. Gassing on the zinc electrode will start, and if the charge is not stopped in time, the process of zinc dendrite formation will start. Every cycle sees the possibility of zinc dendrite formation, since the zinc is fully charged at the end of charge, whereas in the unformed cell, the zinc is charged only to the extent of the formation of the silver electrode. This means that there is a reserve of unformed zinc oxide that acts as a buffer against the formation of zinc dendrite. Eventually, with continuous cycling, the zinc electrode reaches a point where it is fully charged at the end of charge of the cell, but at least this stage has been delayed for a long time.

Limiting Zinc Electrode

Although the cell is frequently designed with excess zinc material, it happens that during the life of the cell, the zinc electrode does not perform as it should, not necessarily because of loss of zinc, but for other reasons, such as a phenomenon called sometimes "passivation" of the zinc metal. This term covers many facets of the imminent failure of zinc, such as densification of the active material, dryness of the electrode (inside and on its surface), and slumping of the zinc at the bottom of the cell, contributing to the thickening of the electrode bottom and the thinning of the top. Upon analysis of the zinc electrode, it is found that the total amount of zinc should have been sufficient to carry the total discharge but could not because of its incipient failure.

This limitation of the zinc electrode can be detected in the discharge curve of the cell. In a healthy cell, starting fresh, the silver electrode is the limiting electrode, which shapes the form of the discharge curve. The curve is flat (in the monoxide portion) until it drops abruptly when the silver oxide is exhausted. Very little capacity remains after the cut-off voltage. In a zinc-limiting cell, the shape of the discharge curve is different in that it shows a slope and reaches the cut-off voltage only gradually and slowly. If such a cell is opened, one can see the positive plates still black in some areas where the opposite electrode areas have zinc-active material missing. Figure 1 is a depiction of the two forms of discharge.

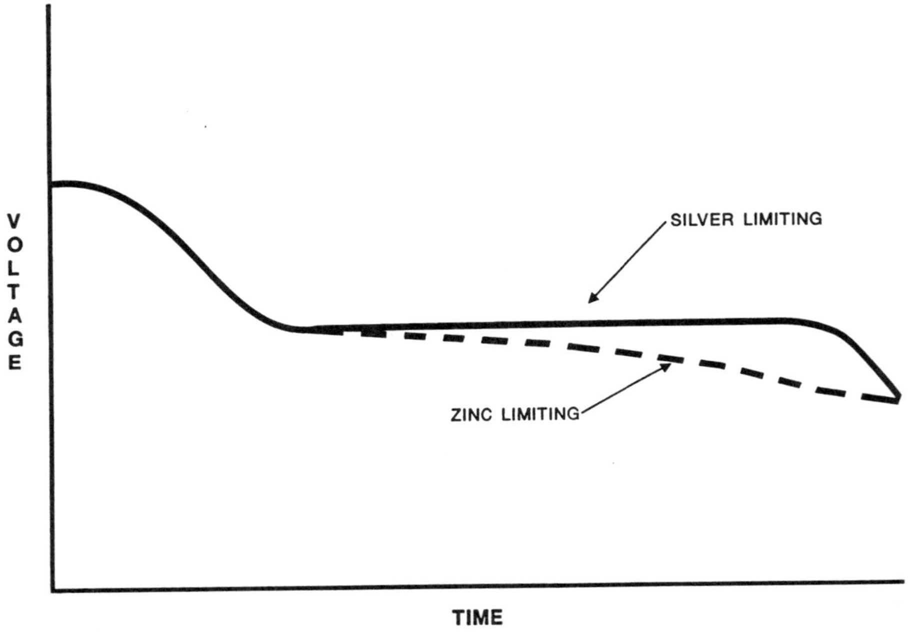

Figure 1. Discharge Curves: Voltage vs. Time for Silver-Limiting and Zinc-Limiting Cells

CHAPTER VI
Separator

As mentioned previously, the preferred separator (and certainly the only viable one to date) is cellophane. It is specially made as battery grade, desulfurized, and glycerin-free. Since this subject is important and certainly determinant in the life of the silver-zinc system, it is being treated separately.

The basic screening methods for characterizing the separator have been treated in detail in a brochure titled "Characteristics of Separators for Alkaline Silver Oxide Zinc Batteries: Screening Methods," edited and published by the former U.S. Air Force Aero Propulsion Laboratory. The methods are still valid today, but a few additions are needed with regard to the quality, reproducibility, and manufacturing specifications of cellophane.

Since present-day cellophane characteristics may deviate from those of the cellophane (procured from Dupont) tested over 20 years ago, it may be worth repeating some of these tests and see how they compare. Also, to update the discussions, a few comments are in order with regard to each of the subjects treated in this brochure and listed below in the same order.

Test Methods Listed in U.S. Air Force Brochure

Dimensional Stablity

Only cellophane is of interest since it is the only material used in today's silver-zinc batteries. The cellophane used is the one obtained from Flexel Inc. Its dry thickness is 1 mil. The stability to be considered is in relation to 45% KOH, not in 31% KOH, as the authors tested it. Also of interest to the battery designer is the data at room temperature (20° C) and possibly at the extreme of the temperature range (55° C), as usually allowed by the battery specifications. The combination of the two parameters used by the authors (31% KOH and 90° C) will degrade cellophane very fast.

It will be worth repeating these measurements in 45% KOH at these two temperatures, as part of acceptance criteria for the present-day cellophane.

In general the designer uses a wet swollen thickness of 3 mils in 45% KOH at room temperature, although 2.8 mils may be the right value on the test.

Tensile Strength

The tensile strength may be useful, but only in the wet state, to establish that no tear is to occur under undue pressure by the electrode dimensional changes after wetting, forming, and cycling of the cells. It may be used as a benchmark for new separators resulting from a modification of cellophane by treatment, also when tested in a new electrolyte different from pure 45% KOH.

The data obtained by the authors and of interest so as to be retained are as follows:

Sample size: 3″ × 2″ Breaking load (lbs)	
Dry	40
Wet	12

Sample size: 5′ × 1″	Transverse	Longitudinal
Dry	15.3	23.4
Wet	1.7	4.5

With the standard methods used by the authors, the breaking load (lbs) was measured for various materials. We will retain only cellophane 300 (which is the 1-mil cellophane used today).

NOTE: The wet samples are described by the authors as being dipped in 40% KOH for 24 hours, then exhaustively washed with distilled water to remove the electrolyte (time unknown). It will be better to use 40% KOH and not use any water, since the cellophane retains a better strength with higher KOH concentration and with no contact with water at any time.

The new cellophane should be tested with this information in mind for comparative purposes, either as is, modified, or in a new electrolyte other than 45% KOH.

Electrolyte Absorption and Retention

The procedure is relatively simple for the measurement of weight of electrolyte absorbed per weight of dry separator (B) and the percentage of electrolyte retained after acceleration (%). Although the authors are careful to note that the concentration of electrolyte to be used should correspond to the concentration of electrolyte used in the actual cell, no concentration number is given as to what was used for various materials tested.

Selecting only the materials of interest to the silver-zinc battery, the data obtained are as follows:

Material	A (mils)	B	%
Woven nylon	3	0.8	72
Nylon felt	6.6	12	93
PVA film	4	2	95
Cellophane	1.5	1.9	100

It is assumed that the thicknesses given are for dry materials. What's being used today is not exactly represented in this table, but we can safely assume the following: the first two are interseparators, used only for the purposes of wetting, electrolyte retention, and separation of the positive plates from the main separator.

The woven nylon is relatively close to the 2 or 3 mil being used today. The nylon felt is close to the nonwoven polyamide (nylon), so-called Pellon, used today in nominal thickness of 6 mils.

The next two are real separators, PVA not widely used and cellophane mainly used as is or modified; however, the values used today are 1-mil-thick PVA and cellophane. However, the values given in the table can easily be converted. A note of caution: the electrolyte has to be 45% KOH and the tests should be repeated with the newer cellophane. The wet thicknesses are:

for 1-mil PVA, no more than 1.5 mils
for 1-mil cellophane, no more than 3 mils

The absorption values are more important than the electrolyte retention after acceleration since the electrolyte remains in the separator and is not ejected at any time. A more important characteristic should be the capillarity figure showing how fast the material absorbs electrolyte from its surroundings. This figure will establish the wicking capability of the materials placed in a tight configuration between electrodes. An improvement of this wicking capability is desirable, since some large cells suffer sometimes from dry spots in the center of the electrodes, thus leading to unformed areas of the electrodes and consequently capacity reduction.

Pore-Size Determination by the Water-Permeability Method

There are too many unfounded assumptions in the method described here to have any value in the practical measurement of the pore size of materials. No data are given for any material. On the other hand, cellophane in water will deteriorate fast and would lead to erroneous results. This method should be disregarded.

Pore-Size Determination by the Electrolyte-Permeability Method

Although it may be successful when using microporous polyvinyl chloride (PVC) separator and 31% KOH, the method has not been applied to cellophane and 45% KOH. No data are reported on materials used in silver-zinc cells.

Electrical Resistance by the Direct Current Method

The method was tested on a film of fiber-strengthened regenerated cellulose material, 2.9 mil thick when dry and 5.7 mil thick when wet, with 40% KOH. The electrical resistance found was for 45% KOH 32 milliohms-in^2.

It is known from other experiments on the 1-mil-thick dry cellophane, swelling to 3 mil when wet with 45% KOH that the resistance value is 10–11 milliohm-in^2.

It seems that the method could be simplified for quick inspection of various lots of cellophane.

Electrical Resistance by the Alternating Current Method

The method appears to yield results in line with known electrical resistance values of cellophane. In 31% KOH, the 1 mil dry cellophane used swells to 2.5 mils and has an electrical resistance of 10 milliohm-in^2. Other materials are not used in silver-zinc cells.

Degradation by Soluble Silver

Soluble silver oxide attack and deposition of silver in cellophane occurs slowly during cycling and even on prolonged charged stand at room temperature. At higher temperatures, this degradation occurs exponentially so that it can be measured in a few days at most and extrapolated to lower temperatures. It has been established that cellophane immersed in hot KOH saturated with silver oxide accumulates metallic silver at an approximately linear rate with time. For a given KOH concentration, a logarithmic plot can then be made of silver content versus the reciprocal of the absolute temperature, which can give the rate at lower temperatures by extrapolation. This can give an estimation of the disintegration time at room temperature (or at least a silver content).

Cellophane is also attacked and degraded in plain KOH without silver oxide, also at a faster rate with higher temperatures and even faster when in presence of oxygen or air. Another disintegration mode is the hydrolytic effect.

For example, it was found that fiber-strengthened regenerated cellulose degrades in 40% KOH saturated with silver oxide at 120° C in 3 hours and in 36 hours in the absence of silver oxide (in anaerobic soaking).

However, no attempt has been made to test cellophane as presently used (1 mil thick). The closest tested is the regenerated cellulose or the same strengthened with fibers (sometimes called fibrous sausage casing). The data obtained on the silver content versus time for each temperature are very close. This material, so-called sausage casing, is not used any longer in the manufacture of silver-zinc cells because of its inherent unreliability and the difficulty in the assembly of cells: it is made in tube configuration and has to be ordered in exactly the proper size so as to fit the electrodes inside the tube. This material is 2.5 to 3 mil thick when dry and swells to 5.7 to 6 mils when immersed in 45% KOH. Extrapolating the data obtained for regenerated cellulose to cellophane may not be completely correct but it may give an idea of the order of magnitude of the silver content.

The author noted that at room temperature the first layer next to the silver plate reached quantities of 60 to 80 mg/in^2 of silver in 800 to 2,000 hours, while total disintegration was observed in 4,000 to 6,000 hours in cells operating at room temperature. One single layer is approximately equivalent to 2 layers of the 1-mil cellophane, which swells to 3 mils. The silver content of 80 mg per in^2 is close to the value of 100 mg found for 2 layers of cellophane. However the empirical observations in actual cells show that when the proper regenerated cellulose is used, it has an advantage in terms of silver degradation over plain cellophane (see next paragraph) even for the same equivalent thickness. It may well be that the cellophane process should be reviewed in this light.

Since this test is important, the test method is described in detail in "Tests Recommended as Most Important," page 55.

Degradation by Oxidants and Alkaline Solutions

The methods (hydrolysis and oxidation) used give some indication of the cellophane characteristics.

The hydrolysis in 31% KOH saturated with zincate in the absence of oxygen gives a breaking load of approximately 16 lbs for a degree of polymerization practically unchanged (450 to 500) over a test period of 80 hours. In the presence of oxygen, the degree of polymerization drops from the original 520 (breaking load 13 lbs) to 461 (breaking load 13 lbs) in 8 hours and 342 (breaking load 8 lbs) in 24 hours.

The oxidation by potassium permanganate, measured by weight loss, is shown to be 69% for the 1-mil cellophane and interestingly only 38% for the fibrous sausage casing (discussed above).

These methods should be reapplied to the present-day cellophane or any new processed cellophane and may be used for inspection during acceptance of the product.

Electrolyte Diffusion

This chapter provides information on the rate of diffusion in cellophane and fibrous sausage casing, which was essentially the same considering the thickness, about 0.0006 moles per minute per square inch, which corresponds to a discharge current density of approximately 1 A/in^2. It was concluded that the potassium hydroxide diffusion through the separators is not a limiting factor in the performance of silver-zinc cells.

Silver Diffusion

The method tries to determine the rate of silver diffusion through the membrane. However, in the case of cellophane, it is known that soluble silver oxide does attack the cellophane, reacting with it while at the same time going through to reach the next layer and so on. It therefore appears that the amount of silver going through is less important than the silver pickup by the membrane. Cellophane can be used as a standard for comparison purposes for other new membranes. A membrane that lets silver go through while not attacked or degraded by silver may be valuable. An argument could be made for a membrane that stops silver by sacrificial degradation, thus preventing silver from going through or at least minimizing its diffusion. Silver-zinc cells that failed by silver-penetration shorts show that the short comes from silver trails through each layer of cellophane. If the membrane lets soluble silver go through without retaining any amount, then shorting by conductive silver through the layers of membranes would be nonexistent.

The author claims that "it does seem likely that a membrane which allows silver to diffuse through at a much greater rate than does cellophane would not be an effective separator because then appreciable amounts of silver could reach the zinc electrodes and cause them to self-discharge." The best zinc electrodes have been required to be fabricated with the use of silver grids and silver current collectors, and no significant self-discharge due to the presence of silver in the zinc electrodes has been observed over the past 30 years.

Zinc Diffusion

The method gives information on the rate of zincate diffusion through the membrane. The author noted, "there seems to be no particular harmful effect of soluble zinc on the positive plate and its distribution throughout the cell would be innocuous, except that as long as it can exist in solution in separator layers and their associated electrolyte, there is a supply of zinc (zincate) which could feed the growth of metallic paths through the separator during charging of a cell." However, the condition of shorting by zinc dendrites exists only when one overcharges the cell past the 2.05-V cut-off voltage limit. As a matter of fact, some manufacturers have been using zincate solution (close to saturation point) as electrolyte to fill the cell in an effort to minimize zinc dissolution from the zinc electrodes.

The author notes also that "there would be an advantage for a separator material to have pore sizes small enough to prevent passage of zincate ions, yet allow free passage of migrating ions of the electrolyte and of water in order to provide conductivity." But the size of zincate ions is of the same order as those that provide conductivity through the pores.

The membrane used in nickel-zinc cells is a good example, since it is desirable to stop the migration of zincate ions, which have a tendency to clog the nickel-active material and reduce its performance. The membrane most commonly used at present is microporous polypropylene sold under the name of Celgard, which does not stop the zincate ions.

For cellophane, the zincate diffusion rate found by the author using the method described, at room temperature in 45% KOH and using 1 M ZnO solution over a period of 115 minutes, is 1.59 micromoles per minute per square inch.

Zinc Penetration

By far, this is the most important test for a membrane intended to be used in silver-zinc cells. Its property to stop zinc dendrites will determine the life of the cell with respect to short by zinc penetration.

When zinc is plated out of alkaline solution, the product is dendritic mainly on sharp protuberances or the sharp edges of the zinc electrode; sometimes it is spongy and soft, depending on the local current density. This uncontrolled growth of zinc will penetrate the membrane (cellophane in this case) and eventually reaches the positive electrode, thus creating an internal electronic short, which may be violent, if sufficient current goes through to generate heat to dry up the separator area and start boiling of the electrolyte and eventually a fire.

Of interest among the materials tested were (wet with 44% KOH):

1. The 1-mil cellophane (swelling to 2.8 mils wet)
2. The silver-treated 1-mil cellophane (swelling to 2.8 mils wet)
3. The 3-mil sausage casing (swelling to 10.8 mils wet)

The zinc-penetration test is related to the time to short given in minutes. The table shows the average time T to short, the zinc-stopping factor ZSF (the quotient of the previous time to a standard taken as the silver-treated cellophane) and the merit of figure MF (defined as the quotient of the zinc-stopping factor by the wet thickness). The merit value (MV) is more properly defined as the time in minutes divided by the wet thickness:

	T	ZSF	MF	MV
		per wet mil		
1. Cellophane	225	0.78	0.28	80
2. Treated cellophane	287	1.00	0.36	102
3. Sausage casing	2087	7.23	0.67	193

From these numbers, it can be seen that 3 layers of cellophane have a better merit value (3 × 80 = 240) for a wet thickness of 3 × 2.8 = 8.4 mils than one layer of sausage casing (193) for a wet thickness of 10.8 mils. Four layers of cellophane have a MV of 320 for a thickness of 11.2 mils, slightly over the sausage-casing thickness (10.8 mils) with a better than 50% in merit. On the other hand, if the sausage casing could be made in 1-mil-thick sheets, like cellophane, its merit value would be more than twice that of cellophane.

Tests Recommended as Most Important

Beside the usual tests to determine electrolyte absorption, tensile strength in a wet condition, and ionic resistivity in 45% KOH, the following tests should give an estimation of the value of any newly improved cellophane in the performance of a silver-zinc battery. Tests 1 and 3 are simplified versions of the methods listed in the U.S. Air Force Aero Propulsion Laboratory booklet on screening methods for separators for silver-zinc batteries. They have been mentioned above. Test 2 is a new one, but very important in determining the ultimate life of a silver-zinc cell.

Degradation by Soluble Silver

The first test gives an estimation of the degree of silver attack in a silver-saturated potassium hydroxide solution at elevated temperature to speed up the degradation of the film. The procedure described in the booklet mentioned above is given here in detail.

Procedure

Samples of the cellophane material are cut 1 inch by 1 inch and placed in a covered beaker containing a solution of 45% potassium hydroxide and maintained at 100° C. The solution contains 2 g of silver dioxide and 2 g of silver monoxide per liter of solution. This will be enough silver oxides to saturate the solution and provide a reservoir of solid silver oxides to feed the solution.

Every hour, samples are removed from the beaker, washed in distilled water, and analyzed for silver content by the Volhard method. Enough samples are used so that some may be carried to disintegration and an average value of time to disintegration determined by visual observation. Several beakers may be used for this purpose.

Disintegration is defined here as that condition in which the sample will not hold together in one piece when agitated in the solution or when holes become evident.

Silver Analysis Procedure

The separator sample is put in a 250-ml beaker and 25 ml of 1:2 nitric acid is added. The beaker is heated to aid dissolution of the silver. About 30 minutes is needed. The resulting solution is titrated with 0.1 N ammonium thiocyanate, using ferric ion as an internal indicator. Five ml of ferric ammonium sulfate solution is added. The solution is titrated to a reddish brown or rust-colored end point.

Calculation

Total grams of silver = ml of 0.1 N ammonium thiocyanate times the silver factor for this solution. The silver factor of ammonium thiocyanate will vary with the normality of the solution and should be calculated every time a new solution is prepared by standardizing against silver nitrate. The silver factor should be close to 0.010777 g Ag per ml of ammonium thiocyanate solution. (In first approximation, it is sufficient to use a factor of 0.01.) The total grams of silver so obtained expresses the silver content per square inch of the separator (if a sample is larger than one square inch, divide by the area).

The silver content picked up should be linear with time and the final disintegration should not take more than 8 hours. Although cellophane is also attacked by the potassium hydroxide, this method will compare various cellophane materials for their resistance to soluble silver attack under the same conditions.

The same test without using soluble silver can determine how much can be attributed to potassium hydroxide by itself.

Silver Penetration Test

This test measures the silver content of each layer of cellophane, starting from the silver electrode, as a function of time, when a mock-up cell is submitted to a continuous float charge at elevated temperature to speed up the process of silver penetration.

This test determines the rapidity of penetration by silver through layers of cellophane and disintegration of the first layer under certain conditions of operation of a silver-zinc cell.

Procedure (see fig. 2)

One charged silver electrode, approximately 2" × 3" (thickness not important), is wrapped in a cellophane U consisting of 3 layers (1-mil cellophane) so that the electrode is placed in one of the branches of the U, the other one being left empty. Two counterelectrodes made of solid nickel sheet to which leads have been attached for current collection are placed, one on each side of the wrapped positive electrode. The whole pack simulates a cell consisting of one positive electrode and two mock-up negative electrodes. The pack is inserted tightly in a cell case with plastic shims as necessary to make it tight. The cell is filled with 45%

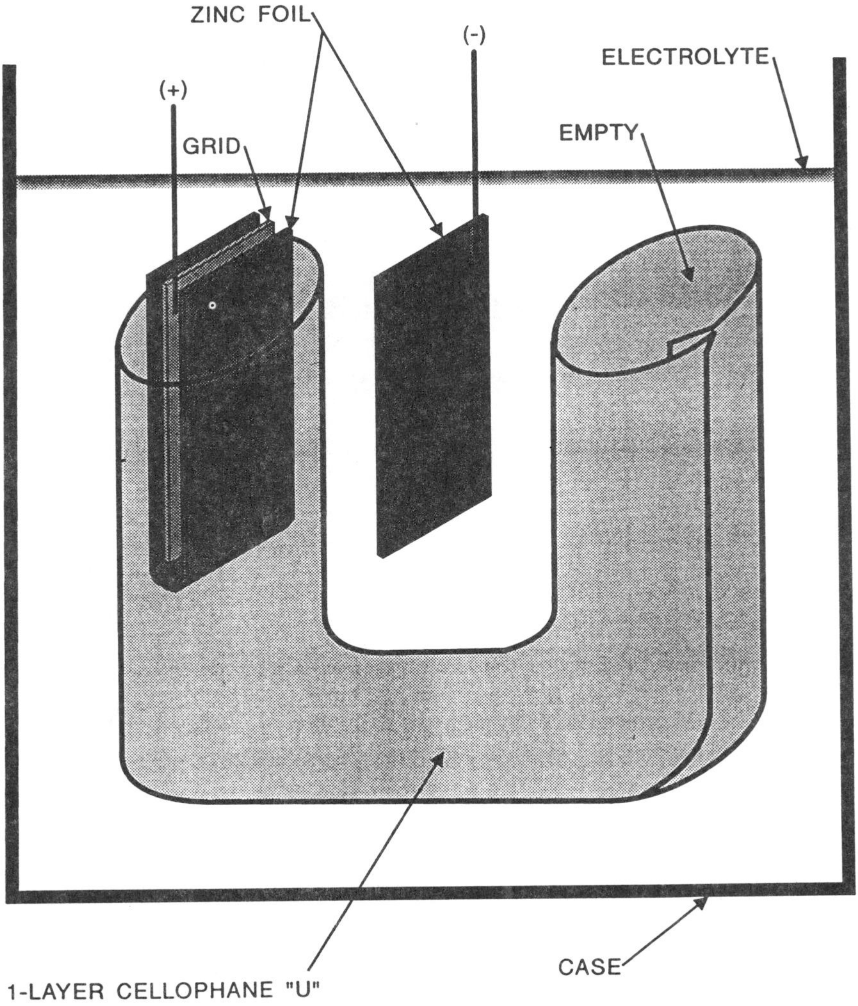

3 Zinc

Figure ~~2~~. ~~Silve~~r Penetration Test

KOH and let soak for one day to ensure full wetting. The cell is submitted to a temperature of 50°C to accelerate the silver penetration and to an overcharge float current of approximately 10 mA/cm^2 of total surface area of the silver electrode. The electrolyte depleted because of evaporation and dissociation by current should be replenished frequently with dilute KOH. Care should be exercised to remove all hydrogen evolved during the overcharge.

Several cells should be assembled and tested in series so that periodically (every 3 days), one could be removed from the circuit, disassembled and examined for analysis of silver content of each layer of separator.

The first layer may be disintegrated, but the last layer should give an idea of how fast silver goes through the 3 layers and how much is stopped by the second layer in a certain period of time.

Zinc Penetration Test

This test measures the time in minutes for a zinc dendrite to form and penetrate one layer of cellophane, when the mock-up cell filled with a zincate alkaline solution and submitted to a continuous overcharge current shows the presence of an internal short circuit (due to the dendrites).

This test will determine the resistance of cellophane or newly improved cellophane to zinc dendrite penetration in terms of time needed to create a short during continuous overcharge.

Procedure (see fig. 3)

The following procedure is a simplified version of the method for zinc penetration in the U.S. Air Force Aero Propulsion Laboratory booklet on screening methods for separators in silver-zinc batteries.

A zinc foil at least 2-mil thick is folded as a U around a silver foil or silver grid collector to which a few silver wires are attached (acting as leads). Press tight to simulate a positive electrode. Another zinc foil will be used as a negative electrode. The dimensions are not important (small size about 2" by 3" is preferred). The "positive" electrode is wrapped in one single layer of cello-

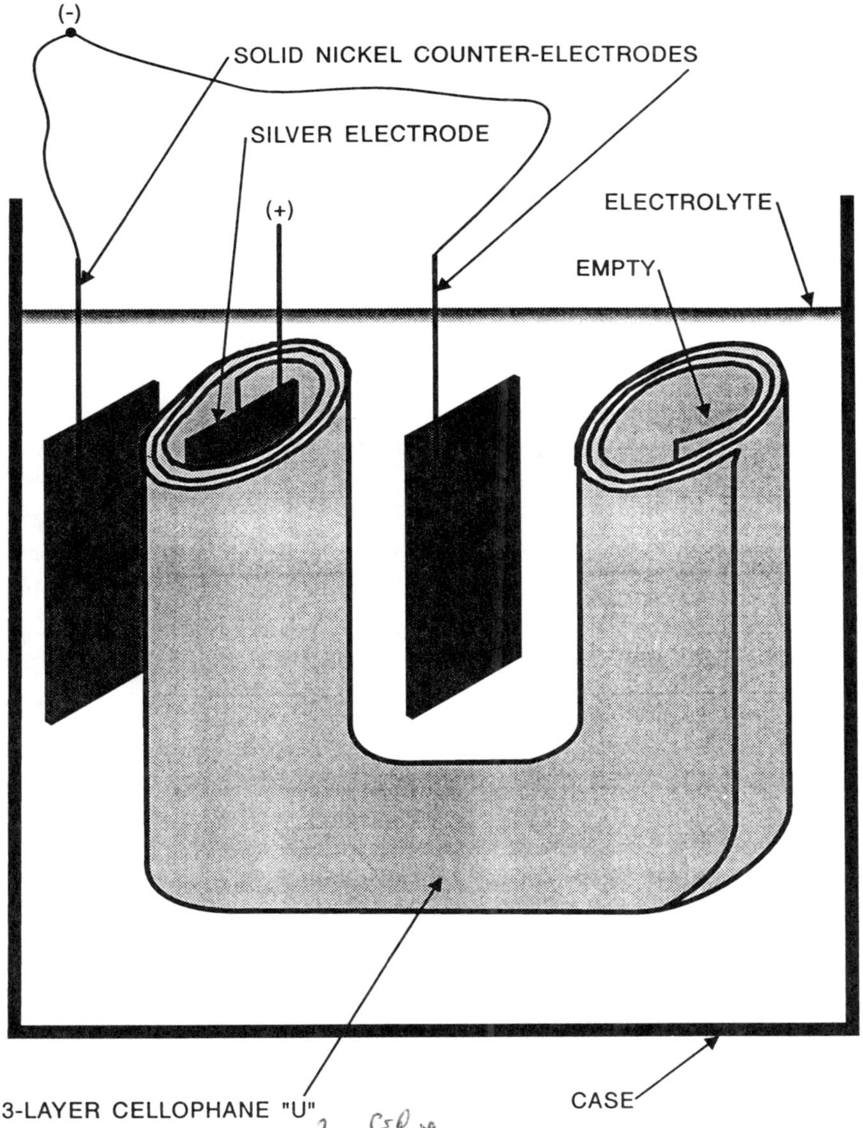

Figure ~~3~~ 2. ~~Zinc~~ Silver Penetration Test

phane as a U, leaving one end of the U empty and the other end occupied by the electrode. The non-overlapping side of the cellophane should be facing the negative electrode so that one single thickness of cellophane stands between the two electrodes. The assembly is inserted in a cell case, tight if necessary with plastic shims. The cell is filled with a saturated zincate solution (zinc oxide dissolved in 45% potassium hydroxide solution) with an amount just sufficient to wet the separator and the plates (no free electrolyte). After a reasonable time for soaking (at least one hour), the cell is placed on continuous float charge at a current equivalent to 10 mA/cm^2 on the basis of one face of the electrode. Zinc will plate out of the solution and deposits on the negative electrode while the zinc of the "positive" electrode goes in solution, replenishing the zincate content. Eventually a trail of zinc will develop from the negative electrode within the cellophane until it reaches the "positive" electrode, thus creating an internal short. The cell voltage is monitored throughout the charge with a very high-resistance voltmeter to avoid drawing current, which could destroy the zinc short when it occurs. An abrupt change in voltage dropping to zero will indicate the presence of a short. The time required to reach this point establishes the value of the resistance of the cellophane-to-zinc penetration.

To give an example of values previously obtained with cellophane types, the 1-mil-thick cellophane gives an average of 225 minutes, the 1-mil-thick silver-treated cellophane an average of 287 minutes, the 3-mil-thick sausage-casing process regenerated cellulose an average of 2,087 minutes.

If standardized to the same wet thickness, considering that the 1-mil cellophane expands to 2.8 mils and the sausage casing to 10.8, the merit-value-per-mil of wet thickness is for:

1-mil dry cellophane	80 per mil wet thickness
1-mil dry silver-treated cellophane	102———
3-mil dry sausage casing	193———

However, it should be noted that 3 layers of the 1-mil cellophane expanding to 3 × 2.8 = 8.4 mils is equivalent to a merit value of 3 × 80 = 240 (compared to the 193 value of the 10.8 wet thickness of the sausage casing). If the sausage-casing process could be made in sheets of 1-mil thickness, the merit value of such a sheet would be twice that of plain cellophane.

CHAPTER VII
Failure Modes

The failure modes of the silver-zinc battery can be reduced to two principal ones:

The first mode is capacity loss where the battery loses capacity progressively with each cycle or on prolonged stand to the point where the remaining capacity is not any longer useful for the intended mission. It is sometimes defined as 80% or 70% of the rated capacity of the battery, not necessarily the original capacity, which is higher than the rated capacity.

The second is the presence of internal shorts, which indicate definitely the end of life of the battery.

Each one has its own symptoms and characteristics.

Capacity Loss

It occurs primarily from the zinc electrode behavior that eventually limits the discharge capacity after a certain number of cycles. The zinc electrode loses its shape (shape change previously discussed elsewhere in this book), the top of the electrode being depleted of zinc while the bottom increases in thickness. Under the best conditions, only the part of the bottom having still zinc material (the bottom) will contribute to the discharge. In most cases, even that portion of the zinc electrode may have more zinc mass than what would be needed for the total capacity, but some of it may be too dense or "passivated" so that only part of it would be used.

The second contributor to capacity loss may be the loss of silver from the silver electrode by silver migration through the electrolyte and the separators, but this portion is very small and does not account for the loss observed. If anything, the silver contribution is due to the fact that silver trails through the several layers of separators finally reach the zinc electrode, thus creating a high-resistance path for internal shorting and thus causing a parasitic loss during charged stand or during operation. A measure of the open-circuit voltage, when the cell is freshly charged, should show that any deviation from the value of 1.86 V at room temperature is the indication of a "soft" silver short.

A capacity loss is sometimes observed during prolonged charged stand. This may be due to a self-discharge, primarily due to the zinc electrode and less from the decomposition of silver dioxide into silver monoxide. Over a long stand, the electrolyte level will be lower than usual, not indicating a loss of electrolyte by evaporation, but rather an equilibrium reached between the levels of electrolyte in the positive compartment and in the negative compartment. In the ordinary assembly of electrodes (normal wrap), the silver electrodes are wrapped in cellophane bags, so that the negative electrodes are placed outside. Therefore in a freshly charged cell, the visible level appears high due to the displacement of water toward the negative electrodes. On discharge, the water returns to the positive side, filling the positive electrode bags and depleting the negative side so that the visible level appears low. Therefore, in a prolonged stand, the top of the electrodes may not be wetted properly and may not deliver all the expected capacity on an immediate discharge after stand. However, after one low rate charge (or even two, depending on the size of the cell), the level of electrolyte will eventually rise and all the electrodes should recover their capacity if they have not been overused.

The phenomenon of capacity loss on prolonged stand is exacerbated when the cell is standing at relatively high temperatures, which may cause an electrode imbalance, since there is capacity loss from the silver oxide electrode and from the zinc electrode self-discharge at a different rate. A recovery can be done with a low-rate charge followed by a mild overcharge, with decreasingly lower currents so as to never exceed the 2.05 V voltage cut-off limit.

The capacity loss is a problem or an issue only when the capacity drops below the mission requirement, which could be 80% or 70% of the rated capacity. This point can be reached sometimes over a relatively small number of cycles, mainly if they are conducted at high rates. Otherwise, the cell can still be used down to 50% in the absence of internal shorts.

Internal Shorts

The most undesirable mode of failure is the end of life by an internal short, which can occur early in the life of the cell, due to a defective assembly, a bad separator, or, more frequently, a lack of proper maintenance, or to an accidental overcharge on a battery where not all cells are monitored for charge voltage limit.

The damage resulting from these conditions is the so-called zinc dendrite formation and zinc penetration, where a trail of metallic zinc goes through the separator layers until it reaches the silver electrode, thus creating an internal short.

Assuming that no defective assembly or bad separator exists, the proper maintenance and operation of the cell should minimize the formation of zinc dendrites.

The maintenance of a cell is essentially the verification of the level of electrolyte, which should be always adjusted if it is too low, only at the end of charge. Otherwise, a low level may lead to drying of the separator and electrode tops, thus creating an area of high resistance and a region of cracked separators so that when the cell is returned to a wet condition, paths are easy for zinc penetration.

The operation of the cell consists in following the procedure in which no cell in the battery is allowed to exceed a certain voltage limit. The common voltage limit is 2.05 V under normal rates of charge for a single cell. However, in a string of cells in a battery, a voltage of 2.02 V is usually recommended, and sometimes, when there are a large number of cells in the battery, a voltage of 2.00 V per cell is generally used. Such batteries are equipped with a voltage scanner to monitor all cells, each in turn very rapidly so that when the voltage limit is reached on any cell, the charge is either cut off or the current reduced to allow the cell to return to a lower voltage. It is worth noting that zinc penetration occurs only if the cell is submitted to a prolonged overcharge. Some batteries in the field, equipped with a voltage-monitoring system, have never failed by zinc penetration.

However, it has been reported that sometimes zinc shorts were observed even when the cell voltages were never allowed to exceed 2.00 V during the charging process. Cells when opened and examined were exhibiting excessive shape change, which may account for creating a nonequipotential surface on the degraded zinc electrode. The top of the plate may be submitted to a high current density while the bottom of the plate is not. The voltage measured may be a mixed voltage of the real voltages of various areas of the electrode. Also the concentration gradient between the top and the bottom of the electrolyte may affect the real voltage reading. This condition, which varies with cycling and age, essentially with zinc shape change, may affect the cut-off voltage at the end of charge. This end voltage will then vary from the time when the cell is relatively new to the time when the cell is relatively aged (implying more shape change). It may be necessary to adjust the cut-off voltage to a different value than at the beginning of the cell life to avoid inadvertent overcharge leading to the formation of zinc dendrites. If the shape change can be monitored as a function of cycling, an empirical formula tying the cut-off voltage to the number of cycles can be established.

CHAPTER VIII
Possible Improvements

The immediate improvements needed in a silver-zinc battery should be concentrated on three specific areas, which are in order of importance the zinc electrode, the electrolyte, and the separator. Although the separator is very important, it is listed last because resolution of the zinc electrode problems (either by itself or in conjunction with a modified electrolyte) will make an immediate impact in the use of the silver-zinc battery in many applications since it will provide a tremendous improvement in its capacity retention with cycling.

Zinc Electrode

The zinc electrode is mainly responsible for the capacity loss and therefore should be improved first, even if the separator is not ideally impervious to internal shorts. The main shortcoming of the silver-zinc battery is its fast capacity drop with cycling. If we could maintain its capacity for a large number of cycles, several applications would immediately benefit, where no other battery could be substituted because of the need of the high energy density and the high rate capability of the silver-zinc battery.

The first needed improvement is prevention of the shape change with cycling. It may not be obtainable by working on the zinc electrode alone, but in conjunction with a modified electrolyte.

One type of electrode that has been used with some success (introduced by General Motors and Lawrence Berkeley Laboratory) consists in the addition of calcium oxide (usually in the range of 15 to 25%) to the zinc oxide. After a few cycles, the zinc and calcium form a complex compound relatively insoluble in the electrolyte. Although this large reduction in zinc oxide may appear to decrease the available capacity of the zinc electrode, it should be remembered that zinc is always in large excess compared to silver by design to compensate for the eventual loss of zinc by shape change. Therefore, the reduction will be compensated by the gain in the capacity maintenance over time. Unfortunately, this type of electrode innovation has not been followed through and should be investigated further.

Instead of mixing the two oxides and making a zinc electrode out of the mix, the mixture could be formed outside the cell in a batch so as to form the zinc-calcium compound prior to the manufacturing of the electrode. If the cell is to contain dry-charge electrodes, the electrodes using this mixture could be formed outside as usually done for dry charge. These two methods will avoid the problem of prolonged formation of the compound in situ (inside the cells), which may require some overcharge, always undesirable in a silver-zinc cell.

Another method is the use of a metal alloy consisting of zinc metal and calcium in the proper or desired ratio. The alloy could be laminated into foils in thicknesses of a few mils (2, 3, or 4 mils are preferred). The electrode can be easily made by folding the foil (previously cut to the proper electrode size) around the grid collector as a "U" with the two vertical sides open and pressing it against the collector for temporary adherence. The total thickness of the foil (or foils) should be calculated in function of the required amount of zinc needed to have the proper excess over silver. The design of the cell-pack thickness should allow for the expansion of the dense zinc metal alloy to the working density of the converted zinc compound (into zinc hydroxide and calcium hydroxide), which would be in the range of 2 to 2.5 g/cc when totally converted. This means an expansion for zinc of approximately 3 to 1. After one or two cycles at a relatively low rate (6- to 10-hour rate), the cell will be usable as if made originally from zinc oxide only.

Other types of inorganic additives for zinc shape retention

have been investigated with no conclusive results, since none has found its way in the silver-zinc batteries used in the field. This approach should be reevaluated in depth and more basically, rather than using trial and error in actual cells and simply testing for life.

Currently an attempt is being made to use an organic compound to retain the zinc particles in place and minimize dissolution in the electrolyte.

In order to maintain zinc electrode shape and its integrity during cycling, it was thought that an electrolyte having low zinc solubility could at least retard zinc dissolution. The preferred electrolyte being 40 to 45% KOH, additives to this electrolyte have been investigated. They may sometimes offer a higher ionic resistance, but when the issue is only longer cycle life, this impediment is not felt important as long as zinc can still behave properly and retain its shape, so as to keep a good capacity retention. A low KOH concentration could fit the requirement of low zinc solubility (while of course having higher resistance), but the cellophane separator will degrade very fast. A new separator capable of resisting in this medium while managing to retard silver penetration would be required.

Electrolyte

Additives to the electrolyte have been investigated several times without significant success. It was thought that the use of dissolved salt or metal in the electrolyte will prevent the dissolution of zinc oxide. Over cycling, it was discovered that zinc oxide gets dissolved in the electrolyte after displacing the alien metal, which precipitates out in the form of hydroxide or other form. Aluminate and silicate forms have been used without success. More recently borate salts are being investigated and first evaluations point to some possible success. In nickel-zinc cells, where a lower KOH concentration is used, an addition of potassium fluoride and carbonate (Lawrence Berkeley Laboratory) appears to hold promise.

There is still much to expect from improvement of electrolyte, an area that has not been fully investigated.

Any new electrolyte (modified KOH or new) should be screen tested by measurement the degree of zinc oxide solubility and the degree of resistance of cellophane to attack, before testing in actual cells. Ionic resistivity should be compared to that of pure KOH, but a slightly higher value should not deter one from using it in actual cells, as pointed out earlier. What is at issue here is the retention of zinc shape and cell capacity first. The internal resistance of the cell is not solely a function of the electrolyte, but of the zinc electrode and of the separators. A cell designer will use several factors to lower the resistance when there is a need, such as lower current density with larger plate areas and higher zinc electrode conductivity with the use of conductivity enchancers such as lead.

Separator

The main component of the silver-zinc cell is the separator that prevents internal shorting by delaying silver penetration and preventing zinc dendrite penetration. It must also resist alkali attack and hydrolysis. Therefore any separator improvement should submit the new material to some screening tests prior to use in actual cells.

The resistance to soluble silver attack and to the electrolyte (KOH with additives or entirely new) should be considered first and compared to that in plain KOH.

The second requirement is its capability to resist silver penetration or to absorb silver with no or minimal degradation. It could also let silver go through without retaining it, so as not to present a conductive trail of silver from layer to layer until silver and zinc electrodes are shorted.

The third requirement should be its capability to prevent zinc dendrite penetration under severe overcharge tests (zinc penetration test). Ideally, if the electrolyte dissolves no or little zinc, zinc dendrite formation should be minimal.

The capability to stop or retard silver penetration can be

mitigated by having a layer of polyvinyl alcohol (PVA) sprayed or coated onto the cellophane on one side or both sides. It needs only be very thin in the order of 0.1 mil to be effective (see reference * below). This will be thinner and more convenient than the use of a full 1-mil film of PVA. For that matter, the same concept can be used for applying a coating of PVA (or spraying) on any material such as the absorbent material used as interseparator (Pellon) or even clogging the pores of a microporous polyethylene film (such as Celgard), since the transfer of ions can go through a regular PVA film. The Celgard has more stability in KOH than cellophane.*

The capability to resist or retard zinc dendrite formation during accidental overcharge is to use a porous nickel coating deposited on a microporous polyethylene film (such as the commercial Celgard), which will act as a quick eliminator of zinc metal (converting it to zinc oxide) when the zinc dendrite comes into contact with any nickel metal.** Although originally intended for the nickel-zinc battery where overcharge has to be done to charge the nickel electrode, the same concept can be applied to the silver-zinc system to a great advantage when combined with other possible improvements.

*See "Invention Disclosures" in the bibliography.
**U.S. Patent no. 4,327,157, Himy et al., "Stabilized Nickel-Zinc Battery."

CHAPTER IX
Related Battery Systems

Silver Systems

There are two silver systems that have been used in various applications, the silver-cadmium and the silver-iron. Both systems use a solution of potassium hydroxide. In both, the negative electrode is stable and does not suffer from the problems of the negative electrode (zinc) of the silver-zinc system, but they are still subject to the same type of silver penetration.

Therefore a separator system capable of retarding silver penetration is still needed.

Silver Cadmium

In this system, the conventional separator system is similar to the one used in silver-zinc: absorbent material as interseparator and cellophane as main separator. Since there is no shape change in the cadmium electrode, the capacity retention is better than that of silver-zinc so that the battery life is somewhat longer. There is still some cadmium solubility in the electrolyte (still 45% KOH because of the use of cellophane) and therefore some slight cadmium penetration can be observed over a long period, no worse than that observed in nickel-cadmium cells. The silver-cadmium cell is capable of 200 to 300 cycles. Since the cadmium electrode does not exhibit any gassing, the system is easy to operate in a sealed condition.

Silver-Iron

This cell is submitted to the same silver penetration, but somewhat at a lower rate, since the electrolyte used is in the range of 31 to 35% KOH. The separator system used has been of the Celgard type. Three layers can provide approximately 200 cycles. However, the iron electrode requires heavy overcharge and therefore can gas profusely, leading to frequent additions of water.

Zinc Systems

Nickel-Zinc

The nickel-zinc system comprises two conflicting electrodes: one, the nickel, requires overcharge for completion of its capacity, and the other, the zinc, is not amenable to overcharge, since it promotes zinc dendrite formation. One way to get around this dilemma is first to overdesign the zinc electrode (with excess zinc oxide active material) so as to retard the effect of overcharge on the charged zinc. Second, the nickel electrode should not be of the sintered plaque type, which consists of an excessive amount of inert porous nickel (for active material retention). Therefore, one way to circumvent this overcharge on the nickel is to use another type of electrode, either the pasted type or the fibrous type (nickel-coated plastic or carbon fibers).

Relative success has been obtained so far with this type of approach. Although the electrolyte is in the range of 35% KOH (with some lithium hydroxide additive as in nickel-cadmium cells), there is still shape change and capacity loss with cycling. To ensure a longer life, the cell should operate in a sealed condition with the possibility of recombining excess oxygen gas generated at the nickel electrode during charge.

The separator system has to contend with zinc penetration. To date, only Celgard has been used, but improvement to this system is certainly required before a nickel-zinc cell can be developed and called successful in terms of long calendar life, long cycle life, and good capacity retention with cycling.

A type of charge that can bring some improvement in the nickel-zinc cell is the use of a pulse charge current or also an asymmetric pulse charge (consisting of periods of charge alternating with short periods of discharge) as described in the patent U.S. 4,327,157 listed in the bibliography of this book.

Recently, a special electrolyte developed by Lawrence Berkeley Laboratory, using a mixture of potassium fluoride and potassium carbonate in a solution of potassium hydroxide, has shown a good improvement in terms of zinc shape change and cycle life in a sealed cell configuration (see reference 4 of the bibliography).

Another promising line has been investigated with the use of calcium hydroxide in the zinc electrode and a low KOH concentration electrolyte (see Adler et al., in reference 1 of the bibliography).

Zinc-Air

The zinc-air system is also a close parent and therefore has the potential for zinc dendrite formation, zinc shape change, and rapid capacity loss with cycling. Beside these problems, the system has to use a bifunctional air electrode (capable of sustaining the discharge and the recharge several times over its life). This type of electrode is still far from achieving the desired life. The separator system has to be resistant to zinc penetration.

CHAPTER X
Closing Remarks

The silver-zinc system is still to this day the only viable system capable of high rate and high energy density, and which can be obtained immediately from different manufacturers in any size as a commercial product without going to extreme lengths for designing, testing, and delivering, with a reasonable schedule.

If the battery life is an issue, an alternate can be the nickel-zinc system, which has a lower energy density, but which, even with the present technology, can easily provide at least 200 cycles. The improvement needed in nickel-zinc is the achievement of over 500 cycles, but for many applications where one is looking for a substitute to the short cycle life of silver-zinc, the nickel-zinc as it stands now can provide at least 3 to 4 times the silver-zinc cycle life. The overall cost of the nickel-zinc battery would be cheaper than that of the silver-zinc battery, since fewer cells would be needed for a given overall battery voltage.

If we assume that a high-density battery is required for the mission, then only a lithium battery can exceed the silver-zinc system and will be acceptable if it is capable of high rate, has a substantially longer cycle and calendar life, can be built in various sizes (sometimes large up to 1000 Ah), and is proven safe. At the present time, no such battery is available on the market.

Bibliography

Adler, T. C., F. R. McLarnon, and E. J. Cairns. "Low Zinc Solubility Electrolytes for Use in Zinc/Nickel Oxide Cells." *J. Electrochem. Society* (February 1993).
Chen, J., F. R. McLarnon, and E. J. Cairns. "Investigation of Low-Solubility Electrodes and Electrolytes in Zinc/Silver Oxide Cells." *J. of Power Sources,* 39 (1992), pg. 333–48.
Gagnon, E. G. "Effect of 10% Weight KOH Electrolyte on the Durability of Zinc/Nickel Oxide Cells containing Zinc Electrodes with Calcium Hydroxide." *J. Electrochem. Society* (November 1991).
McLarnon, F. R., and E. J. Cairns. "The Secondary Alkaline Zinc Electrode." *J. Electrochem. Society* (February 1991).
Salkind, A. J., et al. "Self-Decomposition Processes in Silver Electrodes." *J. Electrochem. Society* (August 1988).

Pertinent Bibliography of Works by A. Himy

Books

Silver-Zinc Battery: Phenomena and Design Principles. New York: Vantage Press, 1986.
Department of the Navy: Document NAVSO P-3676 Navy Primary and Secondary Batteries: Design and Manufacturing Guidelines (September 1991).
Department of the Navy and Westinghouse Electric Corporation: NAVSEA Battery Document NAVSEA AH-300 (July 1993).

Patents Copyrighted by A. Himy

1. U.S. Patent no. 4,084,047: "Stable Alkaline Zinc Electrode," 1978
2. U.S. Patent no. 4,327,157: "Stabilized Nickel-Zinc Battery," 1982
3. U.S. Patent no. 4,192,908: "Mass Transport Separator for Alkaline Nickel-Zinc Cell," 1989.
4. U.S. Patent no. 3,625,772: "Electrode-Separator Combination," 1966.
5. U.S. Patent no. 3,369,937: "Multiplate Battery Construction," 1966.
6. U.S. Patent no. 3,223,555: "Sealed Battery Stacked Electrode Assembly," 1965.
7. U.S. Patent no. 3,207,630: "Electrode Assembly," 1965.
8. U.S. Patent no. 3,173,808: "Hermetically Sealed Battery Cell and Process for Making Same (button cell)," 1965.
9. Invention Disclosures:

    - MTD-91008, 1991, Silver Stopper in Silver-Zinc Batteries
    - MTD-91009, 1991, Silver Penetration Inhibitor
    - MTD-91009, 1992, Silver Stopper in All-Silver Batteries
    - MTD-92005, 1992, Zinc Electrode in Alkaline Batteries
    - MTD-94005, 1994, Integral PVA-Cellophane Separator
    - MTD-94006, 1994, Integral PVA-Celgard Separator